General Preface to the Series

Because it is no longer possible for one textbook to cover the whole field of biology while remaining sufficiently up to date, the Institute of Biology has sponsored this series so that teachers and students can learn about significant developments. The enthusiastic acceptance of 'Studies in Biology' shows that the books are providing authoritative views of biological topics.

The features of the series include the attention given to methods, the selected list of books for further reading and, wherever posssible, suggestions for practical work.

Readers' comments will be welcomed by the Education Officer of the Institute.

1983

Institute of Biology
20 Queensbury Place
London SW7 2DZ

Preface to the Second Edition

This book was first published in 1972. In the original edition I attempted to show where real damage is caused by different types of pollution. Writing as a biologist, I stressed the effects of pollution on living organisms, believing that this was the best way to recognize the real, as opposed to the imaginary, effects of toxic substances in our environment. I hoped in this way to help to enable a rational programme for pollution control to be built.

Since 1972 there have been few major changes in the field of environmental pollution. There has been some continuing improvement in the standards of the air in our cities, as shown by the revised figures in Chapter 2, and our industrial rivers are generally a little cleaner. However, there have been serious incidents of sea pollution from oil, and the risks here seem to get greater. Although pesticides are posing less serious ecological problems in the field, catastrophies like that at Seveso, in Italy, when large areas were poisoned by a deadly chemical from a pesticide factory, show that vigilance is still needed. But on the whole, there is evidence that we are probably beginning to win the battle against pollution, but much more effort will be needed before victory is finally achieved.

Monks Wood, 1979

K. M.

Contents

1 What is Pollution?

1.1 Levels of pollution

Our world is full of poisonous substances. Many of these occur naturally, quite independently of any activity of man. Thus, the vapours from an active volcano may contain so much sulphure that plants cannot grow nearby. Rivers flowing through forests may become deoxygenated because so much natural organic matter is deposited in them, and when this decomposes the results are similar to heavy contamination with raw, man-made, sewage. Mercury, occurring naturally in the ocean, may be concentrated by fish to levels which give concern to public health authorities. But when we consider pollution we usually refer to the presence of toxic materials introduced into our environment by man. This does not mean that only man-made pollution is harmful, though the suddenness of the changes induced by him are often more dramatic than the long-term effects of naturally occurring poisons which we may have come to accept.

I consider that we should only speak of pollution where some effect, probably a harmful one, can be recognized. Modern chemical methods are so sensitive that it is possible to detect traces of dangerous poisons everywhere. Our own bodies, even when we are in perfect health, contain quite substantial amounts of naturally occurring substances which are usually considered as poisons, including arsenic, mercury and other heavy metals. In addition, we have picked up without suffering in health measurable amounts of man-made poisons, like the insecticide DDT and industrial chemicals such as the polychlor biphenyls. Our bodies could be said to be 'polluted', at least by the man-made chemicals, but it is better to treat these traces as contaminants unless they can be shown to be having some characteristic effect.

However, the difference between harmful pollution and harmless contamination is not always a clear one. When a poison is present at a level where acute toxic effects can be recognized, this is clearly a case of gross pollution. We see this in a river receiving a massive amount of raw sewage, or near a brickworks pouring out sulphur and fluoride and killing the vegetation. It is more difficult to be sure that a lower level has no harmful effect. In the case of sewage or sulphur, which soon cease to be toxic, small amounts will usually disappear before any unrecognized damage can occur. However, some other poisons are cumulative. They may either be concentrated in the tissues of living animals or plants, or each minor exposure may have an additive effect. Thus a single exposure to lead, arsenic or DDT may do no recognizable harm, though a measurable amount of the poison may be stored in the body of the victim.

Further exposures may produce an accumulation of the poison to a dangerous level. Chronic exposure to low levels of radiation may act rather differently, but with an apparently similar end result. Here each exposure, even the least intense, does *some* damage, even if this cannot be recognized. It is the summation of such effects which produces serious, and irreversible, results. In the case of the effects of cumulative chemical poisons which are concentrated in our tissues, permanent damage may not be inevitable and may be prevented if the toxic substances are eliminated from the body.

This difference between pollutants which are persistent and those which are not, affects the whole subject of pollution control. The majority of pollutants are not persistent. In most cases all that need be done is to dilute the substance sufficiently so that it is below the level at which it is a recognized poison, and the problem is solved. The diluted poison will then be chemically changed into something harmless. Thus, raw sewage can be discharged far out into the open ocean, and it will break down harmlessly. Toxic gases may be blown up into the upper atmosphere and diluted to a harmless level, where they will soon be changed to substances which lack the original toxicity. It is of course possible for such plans to go wrong. Unusual currents following storms may deposit the sewage onto bathing beaches, and abnormal weather may prevent toxic gases from being dispersed and diluted. But, on the whole, dilution and dispersal is a satisfactory means of controlling the majority of non-persistent pollutants.

The persistent pollutants, often spoken of today with more accuracy than elegance as 'non-biodegradable', pose a very different problem. When they are diluted to harmless levels, they may still remain within our environment, with the possibility that they may be concentrated, perhaps by living organisms, until they are once again present at dangerous levels. For instance, a fish may concentrate some of the organochlorine insecticides from the water in which it swims by a factor of as much as 10 000. But there is a limit to this process, and eventually dilution will always produce conditions where insecticides, for example, are lost from the body and not increased in amount. Also, few of even the most persistent chemicals remain unchanged indefinitely. So the fears of permanent global pollution by man-made chemicals are often overestimated. Nevertheless, these persistent pollutants do present very difficult problems to those who have to deal with them, and future industrial development is likely to produce more rather than fewer substances which fall into this category.

1.2   Population and pollution

Pollution becomes a more serious problem as our population increases, and as our industrialization becomes more intense. Primitive man, living in small numbers, had little adverse effect on his environment. His sewage could be harmlessly absorbed by the rivers, and his smoke soon disappeared into the atmosphere. It was when the population grew and when man came to live in cities that his wastes began to make their impact by poisoning the waters and the air. Then industrial development took place, causing serious damage as poisonous substances were directed by man into the wrong situations. It should be remembered that in many cases man did not create the poisonous chemicals. We have large areas where the soil is made sterile by the presence of high levels of lead, zinc, copper or arsenic. These substances all occurred before they were mined and transported to the industrial plants. In their original location, buried beneath the surface, they usually did little harm, though cases of natural poisoning of the soil and the vegetation are not uncommon, but dispersed by industry the effects were greatly increased. Industry also synthesizes new pollutants, such as organochlorine insecticides and polychlor biphenyls.

Clearly, a great fear is that we shall be unable to contain our pollution. The population of the world is increasing, and is expected to double by the year 2010. So far, industrial development is mainly restricted to a few developed countries, but the whole world hopes (probably vainly) to raise its standard of life to those of Western Europe and North America. More people and more industry will pose great problems of food supply, power and waste disposal. The suggestion that the world's population will double every 20 or 30 years until there is 'standing room only' is clearly nonsense; long before that happens our numbers, like that of any other animal, will be controlled by some other means. This means could be pollution. It could be pestilence. It could be a nuclear holocaust. It is clear that if civilization is to survive, the growth of human population must, in some way, be controlled. This, in the long run, is man's greatest problem. In the meantime, life can only be made tolerable if pollution does not increrase, and if the irreversible degradation of our environment is prevented. But the effects of even the most efficient pollution control will eventually be nullified if the world's population continues to increase at the present rate. Population control, though of such paramount importance, is not the subject of this booklet. We may derive some satisfaction from the fact that, in the late 1970s, Britain's population has remained steady, and growth in most countries, even in the third world, is slowing down. If these trends continue, global disaster may be avoided.

1.3   Biological effects of pollution

I have insisted that the term 'pollution' should only be used when there is actual damage to man or to the environment. Physical damage from air pollution to buildings and to metals can be easily demonstrated, and corrosion by water pollution can also be shown. However, I believe that it is the biological effects of pollution that are of paramount importance. These can often be detected before any physical or chemical effects are easily visible. In many cases it is the danger to human health, an obviously biological parameter, which is the accepted reason for pollution control.

At the beginning of this chapter I distinguished between real pollution and the harmless presence of potentially toxic substances at such low levels that no harm is done, and I showed that this distinction was sometimes difficult to establish. Here the biologist has a contribution to make. I believe that we should try to prevent any pollutant reaching a level where any biological reaction can be demonstrated (e.g. an apparently innocuous but nevertheless definite change in cellular metabolism) even if this reaction has not been shown to be harmful. So often in the past the chronic effects of toxic substances, as of X-rays, have been underestimated. This widens my definition of what I would call pollution, to include the levels of potentially dangerous substances which can be shown to have some recognizable biological effect.

This brings us to the question of standards. I have already mentioned that fish can concentrate some poisonous substances, such as organochlorine insecticides, by a factor of 10 000. Thus, water containing DDT at a level of one part in 100 000 000 would be lethal to some fish, whereas it could be drunk with impunity by man, who could only obtain a hundredth of a milligram for every litre imbibed. To the water authority this water could be treated as 'pure', whereas to the angler it would be seen to be grossly polluted. If we are to be concerned with our whole environment, then we must adopt the higher of these two standards. To implement these at present may not always be practicable, but in the long run the full restoration of all damage should be our goal. The point here is that such damage must be easily demonstrated. At the same time, it is important not to waste our energies on trying to get rid of contaminants where absolutely no biological effects on man, on plants or on animals can be demonstrated.

1.4 Possible global effects

My purpose in this booklet is, then, to concentrate on the biological effects of pollution, but first I must mention some of the alarming possibilities of global damage which man's activities could produce. There is now a growing concern among thinking people about the possible effects of the increasing amounts of CO_2 which are being released through the use of fossil fuels, i.e. oil, coal and natural gas. Since the 1972 edition of this book, this subject has excited much more interest, and is now posing a question mark regarding the place coal may play in world energy supplies during the next few hundred years.

When the deposits of coal and other fossil fuels were being laid down carbon from the atmosphere was being incorporated in vegetation which was then 'fossilized'. This process took many millions of years. Now we are releasing this fossil carbon at a tremendous rate, in the form of CO_2. In the last hundred years, levels of CO_2 in the earth's atmosphere have

Studies in Biology no. 38

The Biology of Pollution

Second Edition

Kenneth Mellanby

C.B.E., Sc.D., F.I.Biol.

Former Director, Monks Wood Experimental Station.
Editor, *Environmental Pollution*

Edward Arnold

First published 1972
by Edward Arnold (Publishers) Limited
41 Bedford Square, London, WC1B 3DQ

Reprinted 1974
Reprinted 1975
Second Edition 1980
Reprinted 1983

British Library Cataloguing in Publication Data

Mellanby, Kenneth
    The biology of pollution. – 2nd ed.
    – (Institute of Biology. Studies in biology;
    no. 38 ISSN 0537-9024).
    1.   Pollution – Environmental aspects
    2.   Pollution – Physiological effect
    I.   Title   II.   Series
    574.5'222   QH545.A1

ISBN 0-7131-2776-7

Printed by Photobooks (Bristol) Ltd.

increased by about 10%. This increase only represents a fraction of the CO_2 released, for a larger amount has been absorbed, as carbonate, by the oceans. It is not known whether the oceans will continue to absorb this gas, or whether atmospheric levels will rise at a faster or a slower rate. Most scientists believe that, if we use fossil fuels more rapidly than at present, CO_2 levels will increase faster, and may double within the next 50 years.

As will be indicated in the next chapter, CO_2 is a normal and essential gas in our atmosphere. All life depends on its presence, for during photosynthesis it is taken up by green plants and supplies, directly or indirectly, all the food for all the animal creation. The present levels of CO_2 are surprisingly low for such an essential substance, being only some 0.033% of air, and it is known that raised levels in glasshouses can make many plants grow quicker and better. So some people have suggested that a rise in CO_2 levels would increase agricultural productivity and make feeding the growing population less difficult. It is possible, even probable, that this would be true, provided that the world's climate remained the same.

Incidentally, it is not only the combustion of fossil fuels which may increase the amount of CO_2 in the atmosphere: burning any vegetation has the same effect. Man has destroyed much of the forest cover of the globe, and in the Brazilian tropics this process is still going on at an increasing speed. Much of the wood is burned, so releasing CO_2, but the removal of the trees also reduces the amount of photosynthesis, so that less CO_2 is being removed from the air. Some investigators think this may be more important than the contribution from fossil fuels; all students of the subject agree that it is significant.

The real concern about increasing levels of CO_2 is that the world's climate may be affected. It is known that high levels of CO_2 in the atmosphere may produce a 'greenhouse-effect'. The air will let the radiation of the sun through, as does the glass in the greenhouse. However, when the energy – the heat – is reradiated from the earth's surface, the greater the amount of CO_2 in the air the greater the proportion of the heat which is trapped; this raises the temperature. It is not known exactly how great any particular rise of temperature will be, and how this will be correlated with the level of CO_2. So there are still many uncertainties.

Many people, particularly those living in colder countries, would welcome a warming of our planet. However, the results could be catastrophic. The polar icecaps might be melted. That floating round the North Pole would not alter the water level, but that in the Antarctica and Greenland has a solid base and could raise the level of the ocean by considerable amounts. A general rise in temperature of 5°C could possibly raise sea levels by 10 m; this would flood most of the capital cities of the world, and much of the most productive agricultural land.

Changes of global temperature would almost certainly have many other world-wide effects. The rainfall pattern would probably be altered, so food-growing areas with plenty of water might become deserts, and dry areas could have increased rainfall. We do not know whether these climatic changes would be harmful or beneficial, but it is thought likely that many countries would be seriously damaged.

It must be realized that we are now moving in the realms of speculation. We do not know for certain how much the levels of atmospheric CO_2 will rise in the future, nor do we know exactly how any rise will affect the climate. Many scientists have said that global changes, without man's help, have occurred in the past: there was an ice age about 12 000 years ago, and the planet has had many hot and cold eras. They suggest that our puny efforts will be negligible in comparison with nature's. Others have said that, because of CO_2, we will be unable to use most of the vast stores of coal under the earth, and so this source of energy will be denied to our descendants.

Those concerned with future energy policy are now aware of these dangers. Global monitoring of CO_2 and of climatic change is now being organized by the World Meteorological Organization. This should allow us to obtain an early warning of any possible effects, and give us time to adapt the pattern of energy use before any catastrophe overwhelms us. I believe that, if we are vigilant, we will be able to avoid disaster. But we should now be developing a world-wide system for collaboration in energy use, so as to put any plan into effect as soon as the results of our monitoring programme suggest that this is necessary.

Some people are concerned lest we should asphyxiate ourselves by reducing the supply of O_2. This essential gas arises from plant photosynthesis, part of the same process which absorbs the CO_2. It has also been suggested that felling forests will significantly reduce the amount of O_2 in the atmosphere, and also that pollution of the oceans will prevent the phytoplankton from making its contribution. Recent calculations suggest that the situation is more satisfactory than had been feared. Even if no O_2 were returned to the air, there is sufficient to last for many hundreds of years. Also, even when trees are removed they are usually replaced by crops which make a similar contribution to the O_2 in the atmosphere. We are unlikely to choke for lack of this gas.

2 Air Pollution

2.1 Atmospheric constituents

The atmosphere of the troposphere – the layer of air supporting all plant and animal life and covering the globe to a height of approximately 10 km – consists of 21% oxygen, 78% nitrogen, and 0.033% carbon dioxide (percentages in dry air). It also includes about 1% of argon, neon, helium and other inert gases which, as far as most living organisms are concerned, act much like nitrogen. Water vapour is present in amounts varying from a fraction of 1% in cold dry air to 3 or 4% during the wet season in the tropics. The moisture of the air is clearly of considerable importance to animals and plants, and its variations have biological effects. In so far as man affects the atmospheric humidity this factor could, if unfavourable, be considered to be a form of pollution. However, except in very local situations (e.g. over-dry air in some centrally-heated buildings, condensation of steam from super-saturated exhaust gases), man's influence is negligible, and for our purposes can be ignored.

Pollution seldom affects the basic constitution of the air, for instance by reducing the O_2 available for respiration. A significant reduction can occur in a confined space like a mine, but not in a room like a crowded lecture theatre with the windows closed. Altitude has a far greater effect. On the top of Ben Nevis the O_2 level is reduced by about a sixth, a fall that is hardly detected by a healthy individual. At greater heights, as in unpressurized aeroplanes, O_2 deficiency can be more serious, and extra O_2 was needed by most of the mountaineers who reached the summit of Mount Everest.

Most pollutants are added, usually in quite small amounts, to normal air. The intensity of pollution can be expressed in various ways. We often speak of the weight of the pollutant in a given volume of air, thus SO_2 may occur as one milligram (or 1000 μg) per cubic metre. This method of expression can be used for both gaseous and solid pollutants. However, for gases it is common also to speak of parts per million (ppm), which means the number of cubic centimetres of the gas in a cubic metre. This may be confusing, as for other pollutants in, for instance, body tissues, one part per million means one milligram per kilogram of body weight. To transform parts per million in the air to mg per cubic metre, the following formula may be used.

$$\text{ppm} = \text{mg m}^{-3} \times \frac{22.4}{\text{molecular weight of pollutant gas}}$$

Thus, normal air has 0.033% CO_2 (molecular weight 44). This can also be expressed as 330 ppm or 660 mg m^{-3}. Similarly, if SO_2 (molecular weight

64) occurs at a level of 0.3 ppm, this can also be expressed as 1 mg m^{-3} or 1000 μg m^{-3}.

The effects of most pollutants depend on their concentration. If this is sufficiently high, acute effects will be seen. As far as man is concerned, levels of emission into the air he breathes are usually well below those where acute effects are produced; places with higher levels of pollution are purposely avoided. Chronic damage may occur, but this may be difficult to recognize, and may only be detected when epidemiological data from large populations are carefully analysed. Some plants are much more susceptible than man, and may show obvious symptoms of phytotoxicity whilst animals appear, at least for a time, to be unharmed. The use of such susceptible plants as indicators of pollution may be useful, particularly as they may integrate the results of exposures at different levels over a period of time. (See p. 14.)

2.2   Fossil fuels – smoke

The burning of fossil fuels has been the most serious cause of air pollution, particularly in densely populated industrial countries like Britain. Coal has been the worst offender, in that fires and furnaces where it is burned discharge enormous numbers of particles of different sizes into the atmosphere. The larger particles make up the dust, the smaller, smoke. The dust is usually so heavy that it is mostly deposited near its source, and until recently as much as 1 kg fell in a year on every square

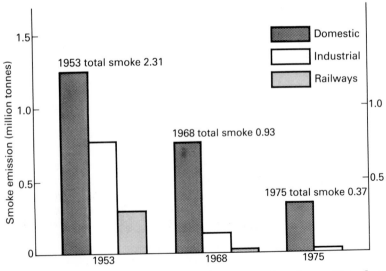

Fig. 2–1   Smoke emissions for the U.K. for 1953, 1968 and 1975. (Data from Warren Spring Laboratory.)

metre near to factories in Britain. It was this dust that made the cities filthy, it helped to make them dark, its very weight harmed the vegetation and choked people breathing it. However, the larger particles were mostly excluded from the lungs by the very processes which made breathing uncomfortable, and the harmful effects were therefore less than might have been expected.

The smaller smoke particles remained longer in the air, and affected larger areas. They were breathed into the lungs, where they remained permanently and blackened the tissues. City pathologists who did many post mortem examinations of corpses became so used to the black lungs of city dwellers, that when they saw the healthy red lungs of a countryman they thought that his tissues were abnormal. Nevertheless, there is little evidence that smoke from coal was seriously damaging, except when it was a constituent, with other toxic substances, of fog. (See p. 13.) Most coal also contained sulphur, and produced considerable amounts of SO_2.

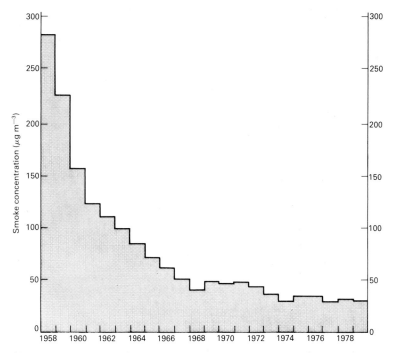

Fig. 2–2   Average smoke concentrations near ground level in London 1958–1975. (Data from Warren Spring Laboratory.)

Oil used as fuel generally produces little smoke and no dust, but some samples are rich in sulphur and produce a great deal of SO_2. Other samples are almost free from sulphur. Natural gas sometimes contains it, but this is almost completely removed before the gas is piped to consumers. Some sulphur-rich oil is also treated and has the sulphur removed before distribution.

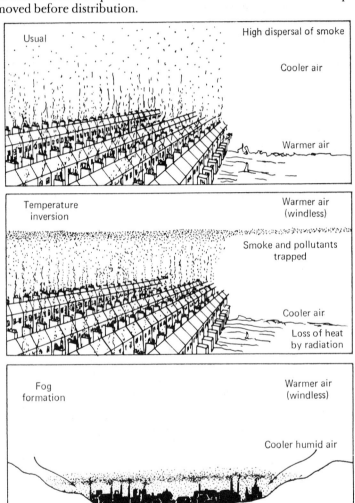

Fig. 2–3   The mechanism of temperature inversion. These illustrations show how change in air temperature can trap pollutants in a dense fog close to the ground. Many industrial towns in Britain are, by their situation, especially exposed to this hazard. (From COMMITTEE OF THE ROYAL COLLEGE OF PHYSICIANS, 1970.)

In Britain, and in most other developed countries, coal use has fallen, and that of oil and gas has increased, during the last 50 years. Thus, in 1910 nearly 270 000 000 tonnes of coal were burned in Britain, and very little other fuel. By 1978, coal use had fallen to less than 120 000 000 tonnes; this is the factor that produced the greatest improvement in air quality in our cities. Figures 2–1 and 2–2 show how, over the last 20 years, smoke levels have continued to fall.

Meteorological conditions can make air pollution less or more serious. When there is a temperature inversion (Fig. 2–3), pollutants may be trapped near to the ground. This will cause a dense fog if smoke is being produced, it will concentrate natural mist, and it will produce 'photochemical smog' (see p. 20) under warm and sunny conditions.

2.3   Fossil fuels – sulphur pollution

Sulphur pollution is a very interesting and complicated subject which deserves particular attention. Nearly 5 000 000 tonnes of SO_2 are discharged into the air in Britain each year. There has been a substantial decrease in emissions in recent years (Fig. 2–4), mainly caused by the use of natural gas and low-sulphur oil. The ground-level concentrations of SO_2, i.e. in the air in which we live, have fallen much more dramatically, being in 1976 only a third as high as in 1957 (Fig. 2–5). This fall has largely been caused by the use of tall chimneys which blow the flue gases high into the air, where they are distributed at lower concentrations over a wider

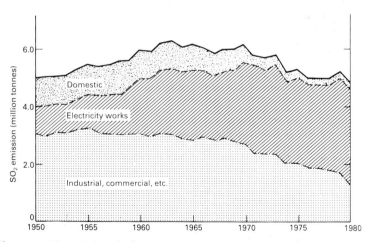

Fig. 2–4 SO_2 emissions in the U.K., 1960–1976. (From reports of Warren Spring Laboratory and the National Society for Clean Air.)

area. This method of getting rid of pollution, by dilution to what were believed to be harmless levels, was generally considered good practice, but in the case of sulphur some doubts as to its wisdom have been raised.

Sulphur differs from many toxic pollutants in that, at the correct level, it is not only not harmful, but it is an element essential to life. Proteins contain 1 atom of sulphur for every 14 atoms of nitrogen. Sulphur is released into the soil and into water from some rocks, and before man appeared there was a natural 'sulphur cycle'. Sulphur compounds were taken up during growth, and released from dying plants (and animals) as the gas hydrogen sulphide. In the atmosphere this was rapidly (in a few hours) transformed to SO_2, some of which was directly absorbed by plants and soil while the rest was oxidized to sulphate which was removed from the air in the rain. The only addition to the cycle came from the rocks, and many soils were so lacking in sulphur that plant growth was poor. The amount of circulating sulphur remained fairly low, as some of the element, usually in the form of insoluble sulphate, got 'locked up' in deposits in the ocean where new rocks were in the process of formation.

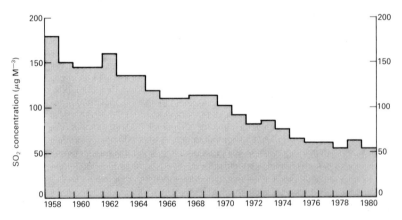

Fig. 2–5   Average SO_2 concentrations near ground level in the U.K., 1958–1976. (Data from Warren Spring Laboratory.)

In excess, all the gaseous sulphur compounds are toxic. Hydrogen sulphide, with its characteristic smell of 'rotten eggs' can be produced by industrial processes and when toxic wastes are deposited carelessly in landfill sites. When man can detect the smell, he is in little danger, but at higher, and potentially lethal, concentrations the olfactory senses are rapidly fatigued and no smell is detected. As mentioned above, hydrogen sulphide is rapidly converted to SO_2, a gas universally present in the atmosphere.

The natural background levels of SO_2 are less than 5 µg m⁻³ which is below the level harmful to any known organism. Mammals appear to be generally more resistant than plants. Experiments with human volunteers show no recognizable effects with levels below 1 ppm (3 mg m⁻³), a concentration seldom found in any British city. Higher levels sometimes occur in industrial plants, where the workers may assert that they relieve the symptoms of nasal congestion accompanying a cold. This apparently beneficial effect should, however, be treated with caution, for it suggests a definite physiological effect which, if prolonged, could easily cause serious damage.

Taken separately, it seems that the main pollutants from burning coal can harm man and his environment, but that this is not often very serious. The main damage, particularly to man, occurs when smoke contributes to the production of fog or 'smog', as happened in London in December 1952. Figure 2–6 shows how the levels of smoke and SO_2 increased during the period 5 to 9 December, and how the deaths more than trebled. This was the result of the weather, when the air was still and there was a

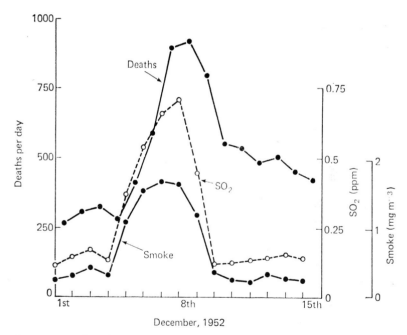

Fig. 2–6   Death and pollution levels in the fog of 5 to 9 December 1952. The graphs show the increased number of deaths during this period and also the rise in the amounts of SO_2 and smoke in the air. The figures for death were provided by the Registrar General's Office. The pollution levels were measured at 12 different stations in London. (From COMMITTEE OF THE ROYAL COLLEGE OF PHYSICIANS, 1970.)

temperature inversion which trapped the pollutants and concentrated them in the lower levels of the atmosphere. The total number of deaths, mainly from bronchitis, pneumonia and other diseases of lungs, and from heart diseases, was, taking the period of the smog and its aftermath, something more than 4000 in excess of the mortality which would have otherwise been expected during the same period.

We do not know exactly what caused these deaths. The levels of SO_2, with a maximum of 0.75 ppm (2000 μg m^{-3}) and of smoke, which reached 1.5 mg m^{-3}, would not themselves seem sufficient to have such serious effects. Other contaminants were present, for example sulphuric acid and oxides of nitrogen, and these are known to be toxic but hardly at such low levels. Man is known to suffer most when it is cold, as it was at this time. The mortality is generally explained as the result of a combination of these various factors.

The death of some 4000 individuals was very serious, but it only corresponded to less than one per 2000 of the population at risk, which makes its investigation difficult. Thus, in a controlled animal experiment with a similar mortality, the numbers used would have to be in the region of 100 000, in both the experimental and control batches, if a significant result were to be probable. It is therefore nor surprising that we are still uncertain of the precise effects of the various pollutants.

There is more evidence of damage to plants, which appear to be much more susceptible than animals: levels of SO_2 between 0.1 and 1 ppm have often been shown to cause obvious symptoms such as leaf blotching and reductions in yield of crops. Coniferous trees grow poorly or even die in many urban or industrial areas. We are now beginning to suspect that crop reductions may occur even when the levels of SO_2 may not be sufficient to cause visible symptoms. The group of plants most thoroughly studied is the lichens, many species of which are absent from urban areas because of their extreme susceptibility to SO_2 damage; they clearly serve as very sensitive indicators. Also, if they are completely prevented from colonizing an area where SO_2 levels are as low as 0.05 ppm, this concentration of the gas is clearly having a biological effect, and it would therefore be rash to state dogmatically that other forms of life, including man, could not be somewhat affected (Fig. 2–7). An authoritative account of the reactions of lichens to air pollution, and their use as monitors, may be found in HAWKSWORTH and ROSE (1976).

In the past it has been usual to consider that SO_2, which is definitely phytotoxic at appropriate levels, is the most damaging form of sulphur to be found in the atmosphere, and to think that when it is oxidized to sulphate it ceases to be a serious pollutant. The situation is much more complicated. In Britain, in areas of substantial atmospheric pollution, damage to plants is usually attributable to high SO_2 levels. As a rule, more than 75% of the sulphur in the atmosphere is present as SO_2 and only the small remainder as sulphate. As the air passes over the countryside, particularly through the foliage of trees, some of the SO_2 is adsorbed and

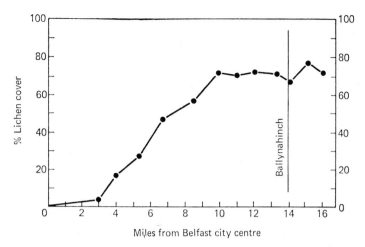

Fig. 2–7 Increase of lichen cover outside the city of Belfast. (After A. F. Fenton, from MELLANBY, 1967.)

removed, but much remains. However, oxidation takes place, and sulphate ions result. This transformation is complex, and may be rapid, under warm moist conditions, when it may be virtually complete in a few hours. In cold or dry air, the oxidation may take many days. The end products are numerous, but sulphuric acid and ammonium sulphate which differ greatly in their toxicity are the most important.

2.3.1 Acid rain

In recent years much concern has been expressed because it is suggested that though in Britain much has been done to reduce sulphur pollution (as shown in Figs 2–4 and 2–5), this is at the expense of Scandinavian countries, for the pollution dispersed by the tall chimneys has been blown over the North Sea to damage their rivers and forests by 'acid rain'.

There seems little doubt that substantial amounts of sulphur from Britain, and from Belgium, France, Germany, Czechoslovakia and Poland, are blown by the prevailing winds to Northern Europe, particularly Scandinavia. It is probable that this passage takes several days, and that what starts as SO_2 ends up as sulphate, some in the form of sulphuric acid. Other pollutants give rise to nitric acid. As a result, the rain falling in Norway and Sweden may have a pH of 4, or occasionally even lower. There is evidence that over the last 20 years the rain has become more acidic.

However, acid rain is not a new phenomenon, in fact some acidity in rain water is normal. Textbooks often say that distilled water has a normal pH of 7, and so it may have, immediately after distillation. However, in the laboratory the distilled water soon comes into equilibrium with the CO_2 in the air, and its pH stabilizes at about 5.7. Rain in completely unpolluted areas has a similar, slightly acid, pH. But in Britain, even in unpolluted areas, the rain is more acidic, and pH levels of 4 or even 3.5 have often been measured. Air in industrial regions is almost always acid. The causes of acid rain are complex. In cities, hydrochloric acid from industrial sources may be present, in addition to nitric and sulphuric acids. In rural areas, sulphur usually makes the greatest contribution.

In Britain, acid rain does little damage. Where sulphur pollution is damaging, SO_2 is the culprit. Britain is fortunate in having basic rocks and rivers with sufficient buffering content to neutralize and render harmless the acid rain, except in some mountainous areas, particularly in Scotland, where the rocks are granitic, the streams unbuffered, and damaging acidity, harmful to fish, may occur.

Why then is there concern in Scandinavia? Total amounts of sulphur deposited there are substantially less than those which fall in Britain. SO_2 levels are far lower, something which may be recognized by the presence in the most 'polluted' areas of abundant lichens. But there is no doubt that, in some streams and lakes, very acid conditions obtain, and that fish in these waters may be harmed or even eliminated. There is concern that forest trees may be damaged, but so far this has not been demonstrated – in fact there is more evidence, at present, of enhanced growth.

Acid rain harms Scandinavian rivers because they are so poorly buffered. Experiments have shown that damage may be avoided if a small amount of lime is added to some streams. Also the weather pattern may make things worse. In mountainous areas at high latitudes all the rain falls for six months or more as snow, and all the pollutants are 'stored' during that period. When spring comes and the snow begins to thaw, it becomes saturated with water. The great bulk of the acid pollutants immediately rise to the top, and come out in the first fraction of the melt. This water may be very acid indeed – specimens equivalent to decinormal sulphuric acid have been found. The ground is still frozen, so this very acid water runs over the surface and forms the bulk of the upland fresh water. Records show that a hundred years ago, when there was little man-made pollution, water from melting snow was acid, and that some fishermen added lime even then, though conditions were much less serious than those found today.

It is difficult to find a solution to this problem. The Scandinavian governments have suggested that Britain, and other industrial countries, should be compelled to control their sulphur emissions. The emitting countries have resisted this, saying that the damage in most areas is slight. Some have suggested sending a few tonnes of lime each spring to neutralize any affected rivers! However, work on removing sulphur from flue gases continues. In the 1950s it was hoped that this could be done economically, as a method of recycling sulphur. Unfortunately, world prices of sulphur fell so much that this became unlikely. But in the future prices may again rise, and if this happens sulphur pollution may become a thing of the past.

Incidentally, sulphur deposition from the air is often a good thing. Many soils are poor in this element, which needs to be added in fertilizers. When the main nitrogenous fertilizer was ammonium sulphate, this was done, often without the farmer realizing what he was doing. Now ammonium nitrate and other compounds containing no sulphur are used; air then makes up the deficit. It is ironical that in Scandinavia the acid rain which damages the fish in a stream may be increasing the yield of crops grown in fields through which the stream runs.

One final point merits mention. The control of smoke may aggravate sulphur pollution. The smoke and other particles absorb some sulphur and remove it from the air. Clean air leaves the sulphur to be blown over the earth's surface.

2.4   Clean air legislation

Most industrial countries, including Britain and the U.S.A., have passed laws to enforce the control of air pollution. In Britain the lethal London smog of 1952 (see p. 13 and Fig. 2–6) aroused public opinion and resulted, in 1956, in the Clean Air Act. This reinforced the controls over industry, and, for the first time, made it possible for local authorities to control smoke from private dwellings. Since then 'smokeless zones' have been set up in which it is forbidden to burn raw coal in ordinary grates, or to burn garden wastes in bonfires. The improvement in air quality in London and many other cities has been remarkable, though in the industrial north of England progress has been slower. Some mining

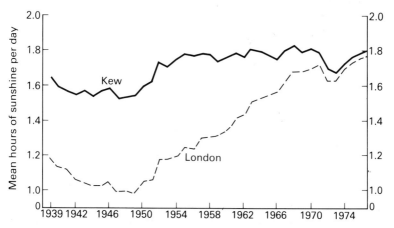

Fig. 2–8   Trend of winter sunshine (December to February) at the London Weather Centre and Kew Observatory. (Data from the London Weather Centre, Meteorological Office.)

communities have refused to comply, and certain public figures have
suggested that coal smoke is 'healthy'. They are guilty of causing some of
their fellow citizens, those with asthma and bronchial complaints, to die
prematurely. The lack of progress in mining communities has its own
economic causes. Miners and their families enjoy a substantial ration of
free, concessionary coal. The Royal Commission on Environmental
Pollution estimated that some 20% of all the air pollution from coal came
from this one source.

The Clean Air Act has undoubtedly been useful, but the main cause of
the improvement of air quality has been the reduction of coal burning,
and its replacement by cheaper and more convenient fuels like gas and
oil. Figure 2–8 is interesting in this connection. It will be seen that since
1970 the hours of winter sunshine in central London have been very
nearly the same as in more rural areas. It should be noticed that the
improvement in London started in 1949, and that the *rate* of
improvement was, if anything, higher in the years before the Clean Air
Act was passed than after. I do not wish to imply that the Act delayed the
process of improvement, I am sure that it did the reverse, but only
because we had the means and they were economically advantageous.

2.5   Pollution from transport

In 1953, railway engines burned coal in Britain, and made a substantial
contribution to air pollution (Fig. 2–1). They are now powered by oil and
electricity, and so produce little direct pollution, though they contribute
to the environmental effects of electricity generation.

Diesel-powered lorries are often seen to belch black smoke through
their exhausts. This is against the law, for a properly maintained diesel
engine is very clean, producing little smoke and almost negligible
amounts of toxic gases.

It is the petrol engine which makes the main contribution to air
pollution. Car exhausts contains substantial amounts of carbon
monoxide, carbon dioxide, oxides of nitrogen, ozone, unburned
hydrocarbons and lead. Under some conditions these substances are
harmless, under others they are deadly. They may have effects may miles
from the place where the vehicles operate.

2.5.1   *Carbon monoxide*

Carbon monoxide is a dangerous poison, causing many deaths a year
when cars are kept running in closed garages. This is an efficient means of
committing suicide. The level in busy city streets has given rise to much
concern, because carbon monoxide has such a high affinity for
haemoglobin in blood, producing carboxyhaemoglobin which interferes
with the transport of O_2. However, surveys in heavy traffic have seldom
shown levels of carboxyhaemoglobin of more than 4%, and this has not
been conclusively shown to have any adverse effect on the behaviour or

the efficiency of the victim. A normal individual is unlikely to be affected, though someone suffering from anaemia might show appreciable symptoms of O_2 deficiency. Much higher levels of carboxyhaemoglobin are commonly found in the blood of cigarette smokers without obvious effects. Concern is expressed lest carbon monoxide levels in cities should rise if the doubling of motor vehicles foretold for Britain in the next 15 years occurs. As most city streets are already saturated with stationary cars in traffic jams, the air can hardly be polluted more in these most polluted areas. The bulk of the additional cars will have to find space in other places. Nevertheless, I do not think that we should accept the present levels of carbon monoxide, even if it cannot be proved that they are doing a great deal of damage. Some priority should therefore be given to engine improvements which cut down emission of this substance.

The possible importance of CO_2 is described on p. 5, and does not warrant further discussion here, except to note that if our worst fears are realized, motor vehicles powered by oil, even if this is produced from coal, will have to be banned.

2.5.2   Oxides of nitrogen

When air, which consists mainly of nitrogen (78%) and oxygen (21%) is heated to above 1100 °C, oxides of nitrogen, particularly nitric oxide (NO) are produced. This happens in car engines. The atmosphere may contain a complex mixture of oxides of nitrogen, generally spoken of in pollution studies as NOx to indicate our uncertainty of the exact constitution of the gases at any moment. One oxide of nitrogen is nitrous oxide (N_2O), a very non-toxic substance, used as an anaesthetic ('laughing gas') and formerly released at children's parties to induce hilarity. Cars never produce N_2O in amounts sufficient to affect man. The main source is from the breakdown of nitrogenous fertilizer in the soil. The gas migrates to the stratosphere, and may contribute to the destruction of ozone there.

Nitric oxide is the main oxide of nitrogen produced by car engines. Levels exceeding 0.1 ppm have been measured in Los Angeles in California, but these are much higher figures than are found generally. At such levels, NO appears to be quite harmless to man and vegetation. The danger is that, in sunlight, it is transformed into nitrogen dioxide (NO_2). This is a much more toxic substance. It is a brown, evil-smelling gas which is harmful to animals and plants. Nevertheless, levels of NO_2 in Britain do not seem to reach danger level by themselves. The trouble is that the NO_2 contributes to the formation of even more dangerous gases.

The first of these is ozone (O_3), which is produced when air containing NO_2 is subjected to sunlight. This reaction is comparativly slow, and maximum O_3 levels may be found tens or even hundreds of miles away from the high traffic densities which were ultimately responsible. Thus, in recent years in Britain there have been O_3 levels sufficiently high to damage sensitive plants in Wales, Northern Ireland and parts of Scotland.

The NO_2 ultimately responsible probably came from the London area and from traffic in continental Europe. Levels of nearly 0.2 ppm have occurred in Britain: these are not harmful to man, or to most farm crops, but cause leaf lesions in sensitive varities of tobacco and peas. We now are beginning to realize that O_3 pollution from cars is, potentially, an important problem in Europe, though not as serious as in the U.S.A.

2.5.3   Photochemical smog

A phenomenon which has caused much concern, especially in California, during the last 30 years is photochemical smog. This has been most damaging in Los Angeles. This form of smog is quite different from the once familiar (though now, we hope, extinct) London smog, which was a mixture of smoke and fog (see p. 13). Photochemical smog contains a mixture of unpleasant gases, including PAN (peroxyacyl nitrates), O_3 and NO_2. It is particularly serious when there is bright sunshine and a temperature inversion, when the pollutants are trapped and concentrated near the gound. PAN, which is phytotoxic and irritant to man at levels in the part-per-hundred-million range (i.e. $1-10$ μdm^3 m^{-3}), is produced mainly by the interaction of O_3 and unburned hydrocarbons. PAN has been reported in several parts of the U.S.A., in Japan, and, occasionally, in Europe.

2.5.4   Lead

Another air pollutant emitted by cars is lead. Tetraethyl lead is added to petrol for reasons of efficiency. One disadvantage of this is that the lead makes control of other pollutants by catalytical methods unpractical, as the lead 'kills' the catalyst. It is also a pollutant in its own right. Vegetation near busy roads may contain as much as 500 ppm (by weight) of lead, and as such it is unsuitable for animal or human food. The pollution seldom stretches far from the road, and serious contamination of crops is uncommon. The most serious cause for concern is the lead which may be given out in aerosol form and is then breathed by city dwellers. So far, surveys have been moderately encouraging. Men who worked in garages in the U.S.A. and who had the maximum exposure, had levels of lead in their blood of 60 μg 100 cm^{-3} in only a handful of cases, and this is below the danger level set by the health authorities. Other city dwellers, including traffic policemen, had even smaller amounts.

Nevertheless, there *is* widespread concern and so much of the petrol sold in the U.S.A. is lead-free; the only disadvantages are that this makes it expensive and more is used to travel a given distance. Lead poisoning can certainly be serious, particularly among children. This complaint was widespread in Britain a hundred years ago, when workers in many industries died and when children suffering from 'pica' which caused them to nibble lead paints flaking from the woodwork of slum houses, poisoned themselves. Although the situation is now much better, there is still an appreciable amount of lead in food (some from the solder used in canned food) and in towns where lead water pipes remain. It is estimated that, even in areas of heavy motor traffic, between 75 and 90% of lead entering any human body comes from food and water, and only the small balance from car exhausts. There is some disagreement as to these amounts, some people insisting that lead breathed into the lungs is much more efficiently absorbed into the tissues than is that taken into the stomach. There is also disagreement as to the levels in the human body which are harmful. Government policy in Britain is to reduce the tetraethyl lead in petrol as rapidly as possible, even though the majority of informed scientists think that this is unnecessary. It is also feared that new precautions may introduce new hazards. Some samples of lead-free petrol contain benzene, which could be a serious danger to those who handle it.

The U.S.A. and several other countries are enforcing measures to prevent the more serious forms of vehicle pollution. Engines are being redesigned to lower the emission of toxic substances, or are being fitted with appliances to the exhausts to contain them. In Britain, where the problem is clearly less serious, mainly because on this small island surrounded by sea the pollutants are more easily dispersed, such regulations have not yet been implemented. Also, the cars in Britain are usually much smaller.

2.6   The ozone layer

So far we have been considering ozone as a dangerous man-made pollutant. However, this same gas makes life on earth for man and similar organisms possible. Much of the radiation from the sun is dangerous, and the surface of the earth is protected by the 'ozone layer' in the stratosphere which filters out radiation which would otherwise be harmful. If this layer were destroyed, or depleted, life on earth would be endangered. Even a small amount of O_3 destruction could make skin cancer more common.

Some scientists fear that certain types of pollution will reduce the efficacy of the O_3 layer. Supersonic aircraft flying in the stratosphere give out exhaust gases, including water vapour and oxides of nitrogen, which may react with O_3 and destroy it. Persistent chemicals, 'freons', used as aerosol propellants for hair lacquer, and paints and insecticides, may reach the stratosphere and destroy some O_3. N_2O from cars, and from chemical nitrogenous fertilizers used on farms, may possibly destroy some more. The majority of those who have studied these problems are not greatly concerned, but there is some disagreement in the scientific community. As yet there is no measurable effect of the current use of high flying aircraft and aerosol propellants. Calculations suggest that the rate of O_3 destruction, if it occurs at all, will be slow, and the risk of skin cancer will increase only by the amount naturally incurred by, for example, moving from Birmingham to Brighton in England. Also, those who deliberately expose their skin to the sun are always at some risk, which could be avoided by keeping the body covered. Amounts of O_3 in the stratosphere fluctuate by many per cent over quite short periods, so any change induced by man may be difficult to identify. Nevertheless, although at present there are so many uncertainties, all agree that this is a subject for further research and monitoring.

2.7   The Alkali Act and the Alkali and Clean Air Inspectorate

Today smoke, SO_2 and the pollution from vehicles cause most concern to the public. Formerly, most damage appeared to be caused by emissions from industrial sources. The first Alkali Act was passed in 1867 primarily to control the emission of hydrochloric acid from alkali works. There are still many areas around the sites of old factories and smelters where the ground is made sterile by the accumulation of old pollution by copper, zinc and arsenic. There has been an undoubted improvement in Britain, and for this the Alkali and Clean Air Inspectorate deserves much of the credit. This independent, though government-organized body, has the unusual duty of controlling specified pollution including that caused by metals and fluorine by 'the best practical means'. This system is often criticized by those who think we should have rigid, legal minimum levels to be permitted for each pollutant. There is little doubt that had this course been adopted from the outset, Britain would today be much dirtier than it is, because standards, to be universally applicable, would have been set for the worst areas and would have been lower than those enforceable in the best. Also, there would have been a tendency to try to keep emissions just below the prohibited level. There have been times when the Inspectorate have been criticized for not being tough enough and for permitting old factories to continue to pollute the air, but the total end result has been a lesson to other countries with more rigid but less easily enforceable schemes. Old factories go out of use, and are then replaced by those with higher standards, designed to fulfil the more stringent requirements of the Inspectorate. This ensures a steady improvement of environmental quality.

Within the walls of some factories, higher levels of pollution sometimes continue. Here, levels (as for NO_2) may be monitored to prevent damage to workers. Masks may be worn to protect them from such substances as asbestos, the dust of which could otherwise produce serious bronchial illness and the possibility of lung cancer.

Some asbestos dust is carried by the air in cities, probably as the result of the wear of car brakes. It has been suggested that this dust presents a health hazard, though existing evidence suggests that it never reaches the threshold of the danger level.

The most serious industrial pollution comes from brickworks and aluminium smelters. SO_2 and fluorine from older brickworks with insufficiently high chimneys may damage crops in the vicinity. Newer works dispose of the gases higher and more efficiently.

Copper, arsenic, zinc, lead, cadmium and other metals have, in the past, been belched out by factories, and have poisoned the soil in their vicinity. Grass and other plants have often been killed. However, most plants are very variable, and contain a small percentage of individuals which, naturally, are less easily damaged by, for instance, toxic levels of copper. These resistant plants survive, and eventually cover the polluted land. Scientists are now seeking such resistant 'races' of grass, etc., to be used to revegetate polluted spoil heaps and other areas damaged by uncontrolled pollution in the past.

Fluoride, even at levels of 0.1 ppm or lower, can cause serious damage to plants. Thus, in Norway aluminium smelters at the bottom of deep valleys may be surrounded by many square kilometres of dead and dying coniferous trees. In Britain there have been reports of fluoride damage to agricultural crops, but the most serious effects have been on cattle. When fluoride is deposited on pasture, the grass concentrates the pollutant, and when it is eaten the animals are affected. Mild poisoning mottles the teeth, while in severe cases the skeletal bones are softened and eventually the animals die. Most modern smelters have very high chimneys so this local damage is prevented. But near brickworks fluoride poisoning can still be detected, and this danger prevents cattle being kept in some areas. This type of damage seldom occurs at more than 10 km from its source.

2.8   Tobacco smoking as pollution

It is clear that most other types of air pollution pale into insignificance before the one really serious form – cigarette smoke. There is now no doubt that much lung cancer and various bronchial and cardiac illnesses are associated with smoking. As has already been mentioned, higher levels of carbon monoxide are found in the blood of a smoker than in those exposed to the highest intensities of traffic fumes. The amount of particulate matter tolerated (often unwillingly) by the non-smoker surrounded by smokers is often far greater than that found near the smokiest coal fire – and these are illegal in many parts of Britain. The particulate matter actually inhaled by smokers is of a different order of magnitude from most other objectionable combustions. Although the relationship between smoking and many forms of illness is clearly established, the actual mechanism causing this morbidity is still not fully understood. Levels of cancer differ in rural and urban areas, in different countries and between individuals of different nationalities. Nevertheless, the general picture is quite clear. It is surprising to find that some of those who attack air pollution, perhaps from motor cars, with such vigour, are still prepared to endanger their own health and irritate their associates by continuing with this much more dangerous practice.

3 Water Pollution

Many parts of the inland waters of Britain suffer from gross pollution, by sewage, industrial effluents and the effects of agriculture. The British Department of the Environment divides rivers into four classes, the first being 'unpolluted' and the fourth 'grossly polluted'. In 1975 of the 36 123 km of rivers surveyed, only 1178 fell into class 4, which at first seems satisfactory. Unfortunately, the polluted stretches include most of the lower parts of the larger rivers, where much of the population of Britain lives. This means that a much greater volume of the flowing water is polluted than the figures would, at a glance, indicate. Nevertheless, over the last ten years there has been some improvement. Nearly a 1000 km of previously polluted water has been restored to class 1, and the amount of grossly polluted river has been reduced by a similar amount.

3.1 Oxygen levels

However, many of the rivers which are considered to be 'unpolluted' are in fact suffering from the results of man's activities, and they are, from the ecological point of view, degraded. I have already said that we should only use the term pollution when some harmful effect, often a biological effect, can be shown. A change in the flora and fauna is such an effect, and if the change is for the worse, then this is surely justly described as a symptom of pollution. Even a moderately polluted river can be rendered safe for human use; such rivers are sometimes classified as 'clean' by the water industry.

Air, no matter how heavily polluted, still retains the normal proportions of its major constituents, nitrogen and O_2, and so will almost always contain enough O_2 to support life. Polluted water, on the other hand, may be totally deoxygenated, so that most aquatic forms of life may be unable to exist in it. At best, water only contains, in solution, rather a low level of O_2. Thus at 20 °C a litre of saturated water only contains some 6 ml^3 of O_2 while a litre of air contains over 200 ml^3. Even an unpolluted stream or pond is probably only saturated during the day when plants are engaged in photosynthesis and liberating O_2; at night, when the plants, like the animals, are using it up, the O_2 levels may fall substantially.

The warmer the water, the less O_2 it is able to hold in solution. Thus lack of O_2 is one of the factors in thermal pollution (see Chapter 4). The difficulty for fish and invertebrate life is that the higher the temperature within the range where life is possible, the greater is the need for O_2, and the lower the supply. This is a reason why a change in the temperature of a river or lake may alter the whole balance and range of species present. It should also be noted that water containing toxic pollutants usually has a lower O_2 level. This means that fish must breathe very large volumes of water to obtain their O_2. The water is taken in through the mouth and passed out over the gills. These absorb the O_2; they may also absorb the poisonous pollutants. So the lower the level of dissolved O_2, the higher the dose of pollutant likely to be absorbed.

The commonest type of water pollution is by organic matter such as sewage. This has the effect of stimulating bacterial and fungal growth, and these processes absorb O_2 and so deoxygenate the water. The organic matter is usually measured by its so-called biochemical oxygen demand (BOD). This is based on a test where the ability of polluted water to absorb O_2 is measured – the greater the O_2 demand, the more polluted the sample. The relationship between BOD and the O_2 level in a river, just below an outfall where sewage or other organic matter enters it, is shown in Fig. 3–1. This example shows the chemical, physical and biological effects of a moderately severe pollution, but one in which the self-purifying processes are sufficient to overcome the pollution so that the river is eventually restored to the conditions which existed above the outfall.

The effects of organic pollution clearly depend on the amount of organic matter discharged into a river, and the volume of the clean water present to dilute it. In an extreme case BOD will remain high, and the O_2 level low, and there will be no recovery before further pollution occurs. Then it will only contain bacteria, sewage fungus and particularly resistant animals like *Tubifex* worms, which have haemoglobin in their bodies and which are able to exist in almost O_2-free water. If only a little pollution is discharged into a large volume of clean water, then little modification of the flora and fauna may occur. Intermediate amounts of pollution will produce intermediate results. The most commonly observed effect is a change in the types of plants and in the species of fish. A very dirty river has no fish. The most resistant fish appears to be the eel, the least the trout. In fact trout are excellent indicators of pollution; if they can exist, the water will be fit for almost any purpose.

3.2   Pollution by pathogens

It should be noted that we have so far been considering what can be called 'ecological pollution', with a change in the balance of plant and animal life. For man, particularly when we are considering the field of public health, there may be other considerations. In the tropics we may find water which shows none of the ecological changes characteristc of pollution but which is infested with pathogenic organisms, including the free-living stages of helminths which parasitize man and domestic animals. Here man is simply playing his part as an intermediate host in the complicated life history of the species concerned. The excrement which carried the parasites into the water has been absorbed and has

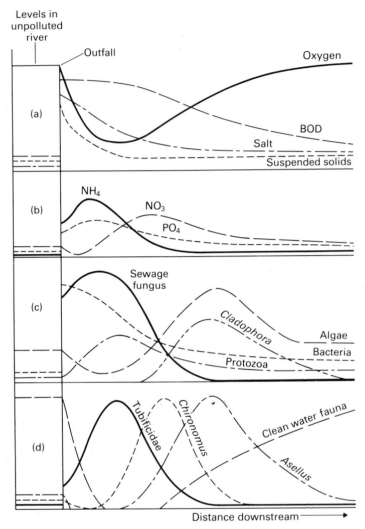

Fig. 3–1 Diagrammatic presentation of the effects of an organic effluent on a river and the changes as one passes downstream from the outfall. (a) and (b) physical and chemical changes. Most recently installed plants tend to reduce ammonia (NM$_4$) and phosphate (PO$_4$) levels, and increase nitrate (NO$_3$) levels. (c) Changes in microorganisms. (d) Changes in larger animals. (From HYNES, 1960.)

helped to nourish the indigenous flora without upsetting its composition. To avoid infection with such parasites, man may alter the whole economy of the river, for instance by canalizing it so that all the flora and fauna are practically eliminated, or by using chemicals which alter the entire ecological balance.

3.3   Pollution by sewage

Today, most domestic sewage in Britain is treated before discharge into inland waters. The bulk of the organic matter is removed, mainly by sedimentation and the action of aerobic microorganisms which act on it as the sewage in suspension passes through filter beds. Unfortunately, in many British cities the population has outgrown the capacity of the sewage works, and so treatment is incomplete. Some completely raw sewage is also discharged. However, the situation is generally improving, and it is likely that in a few years time uniformly high standards of what is known as 'secondary treatment', with the removal of the organic matter and the production of effluents with very low BODs, will prevail. This will be a great improvement, but it will not restore the rivers to their pristine purity, and problems of unwanted algal growth in water-storage reservoirs may even get worse. This is because, in removing the organic matter, increasing amounts of nutrient salts are liberated, particularly phosphates, into the otherwise purified effluents. This gives rise to 'eutrophication', a problem discussed below (p. 30).

Nevertheless, there are substances in domestic sewage which, at present, pass through the treatment plant and pollute the effluent. The most notorious are the synthetic detergents. A few years ago many British rivers were disfigured by masses of white foam sometimes known as 'detergent swans'. This was caused by the so-called 'hard' detergents which were in general use for domestic washing and by the textile industry. As little as 1.0 ppm of these substances caused foaming in the rivers, and reduced the uptake of O_2 considerably. They did not appear to be particularly toxic, and coarse fish continued to be caught by fishing through the holes in the foam. This problem has now considerably abated, as detergents with a slightly different chemical composition which are more easily decomposed by bacteria in the sewage works are substituted. However, some industries still seem to use the resistant substances, so local pollution continues.

Today, the main complaint regarding detergents is that they release large amounts of phosphates which pass into the effluents (Fig. 3–2). In the U.S.A., some states forbid the use of phosphate-rich detergents, though those substituted may not be as efficient in washing clothes. Some new washing powders low in phosphate have presented fresh dangers, including greater toxicity and carcinogenicity. Many British scientists believe that it is best to continue to use the most efficient, phosphate-rich washing powders, and to 'strip' phosphate from sewage effluent. This has the advantage that phosphate from all sources (for by no means all comes from detergent) is removed. Incidentally, it seems that the so-called 'biological' washing powders, containing enzymes, do not present a serious pollution problem. They may be dangerous to the factory workers who produce them, and cases of dermatitis among housewives using

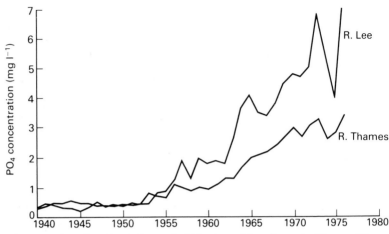

Fig. 3–2 Phosphate levels in two British rivers. (Data from the Thames Water Authority.)

them have been reported, but the active ingredients seem to be broken down in the sewage works.

Domestic sewage is thus seldom acutely poisonous; its harmful effects are the encouragement of the wrong sorts or organism in water enriched by sewage effluent. Industry, on the other hand, may discharge highly poisonous substances. We often hear of fish killed from discharges of cyanide, but damage from metals such as copper, zinc, lead and mercury is not uncommon. The main trouble with some of these metals is that their effects may be cumulative, so repeated exposures to low levels may lead to concentration in the tissues of fish and, at perhaps higher levels, in the birds and mammals which live on them.

Organisms differ greatly in their reactions to these metals. Thus, algae are particularly susceptible to copper, and a level of 0.5 ppm may be used to keep a pond or a reservoir clear. Fish can usually survive exposure to twice this level, though this margin of safety is a narrow one. The long-term effects of sublethal exposure to many of these substances, particularly on the interrelation of different organisms, requires further study. A fauna, almost completely different from that found in nearby rivers with lower levels of heavy metals, is found in Wales and Cornwall in areas in which the rivers are naturally contaminated. Trout and salmon are generally the first fish to be eliminated.

3.4 Regional water authorities

Since April 1974, England has been divided between nine Regional Water Authorities each with the responsibility for a main river basin or catchment area. These Authorities have the functions of extracting, purifying and supplying drinking water in their region, and in treating the sewage in the same area. They are responsible for the control of all discharges into their rivers – both discharges from (their own) sewage plants and from industrial and other enterprises. Before anyone can discharge any sort of effluent into a river, he must first obtain a 'Consent' from the River Authority.

In the comprehensive British Control of Pollution Act, passed in 1974, legislation regarding discharges to rivers was altered. Before that date, although in practice there had been an improvement in many rivers, there was some concern among members of the public because of the secrecy surrounding effluent discharge. The provisions of the new act were all finally brought into effect in 1979. Now River Authorities have the responsibility of maintaining and improving the purity of their rivers, but in a practical way which will not make impossible demands on public authorities or on industry.

For all rivers there are River Quality Objectives (RQOs) which are pragmatically arrived at; all rivers, it is hoped, will become cleaner, but it is realized that this will take time, for some are still very dirty. Consents, for those who wish to discharge effluents, are set out in detail, and are devised so as to at least maintain existing standards. A new factor is that all the details of RQOs and Consents are now made public, and any member of the public who believes that industrialists or River Authorities are not maintaining their standards, has the power to prosecute the offender. It is hoped that RQOs will be raised quite rapidly, and that this new legislation, and the powers of the River Authorities, will ensure the rapid improvement of all fresh water in the near future.

The River Authorities have a colossal task in maintaining water supplies to domestic consumers and industry. Much of the water is abstracted from the rivers, and here there are River Quality Specifications (RQSs) which the Authority strives to maintain; the use of the water is governed by the specification. In dry weather, many rivers now consist largely of sewage effluent. This is, of course, treated before it is distributed, but the effluent must be of a fairly high standard if treatment is to be successful.

We owe much of the improvement of British rivers to the people who fish in them. Fish are excellent biological indicators of water pollution. Under the Common Law of England a riparian owner has the right to insist that only clean water flows past his land, and he can prosecute those who pollute this water. This can be effective in reasonably clean waters, but gross pollution usually has a multitude of sources, so the responsibility of any one polluter is difficult, or impossible, to establish.

As already mentioned, there have been considerable improvements in many British rivers in the last 20 years. Thus the Thames, which was a famous salmon river in the Middle Ages, became grossly polluted in the mid-nineteenth century. The most serious cause of pollution was an improvement in the hygiene of London. Unhygienic earth privies were replaced by water closets, but the ordure was flushed untreated into the river which became almost totally anaerobic, and up to 1957 was virtually free from fish. Then its quality improved and fish began to penetrate – freshwater species from the cleaner water upstream, and marine or euryhaline species from the estuary. In 1964, no less than 41 different species of fish were recovered, and by 1978 this number had increased to well over 200.

3.5   Pollution by farm-animal wastes

In the past, farm animals in Britain lived mostly outdoors. The return to the soil of the excreta was the main method of sustaining soil fertility. Today, an increasing proportion of cattle, pigs, poultry and even sheep, are confined for all, or a substantial part, of their lives to buildings. Many of the farmers rearing these livestock have no arable land on which to use their manure, nor, if they have the land, do they have the labour with which to spread it. Furthermore, in the West of England most farms keep livestock, while there are many wholly arable farms in East Anglia. It would be too expensive to transport farmyard manure right across the country. Thus, much of the excreta has become a source of pollution instead of a valuable resource. Animal excrement pollutes in exactly the same way as human sewage, except that there may be more of it. Thus a cow may produce the equivalent of ten men, a pig of three, and ten hens are equivalent to one man. In 1977 in Britain there were 14 899 000 cattle, 28 030 000 sheep, 7 665 000 pigs and 137 434 000 poultry, so their potential contribution to pollution was more than three times that of the human population which is already grossly overloading parts of the existing sewage machinery.

At one time in Britain there was a substantial government subsidy on chemical manures, which acted as a 'negative subsidy' on farmyard manure, and reduced its use even more. However, the rise in the 1970s of the cost of oil, which is used in the synthesis of nitrogenous fertilizer, has greatly raised the price of chemical fertilizers. This has made it advantageous to use farmyard manure, and it is probable that this trend will continue. If it does, pollution by animal wastes is likely to decrease.

Today, an increasing amount of silage is made, to conserve grass and other crops for use indoors and in winter. Nutritionally this is an excellent practice, but unfortunately silage also produces an effluent which has an immediate effect in polluting water into which it escapes. As more and more silage continues to be made, accidents may become frequent.

3.6   Eutrophication

Where animal wastes are disposed of into ordinary sewage plants, the organic matter is retained, but a substantial amount of the nutrient salts is passed out into the effluent, and contributes to the problem of eutrophication, in the same way as does urban sewage. Eutrophication, in which water becomes more heavily loaded with nutrient salts, is a natural process. Upland rivers and lakes are usually 'oligotrophic'; they contain few nutrients in solution. They remain clear, as plant growth is not encouraged, and fish like trout, which feed largely on insects and other forms of animal life nourished outside the water, flourish. Such waters naturally become eutrophic as they flow into the lowlands, being enriched by salts leached from the land through which they flow. A young lake is oligotrophic; as it gets older, nutrients accumulate in it and it becomes eutrophic. Under natural conditions eutrophication gives rise to a balance, detected in the difference between the flora and fauna of a lowland river or lake and those of the uplands. Problems arise because, where eutrophication is caused by man, the time scale is so reduced that this balance is upset.

Sometimes, as in fish farming, eutrophication is deliberate. The intention is to maximize production, with an interaction between the plant and the animal life, and a surplus production of fish which can be cropped. Considerable skill is needed to preserve a balance. Unintentional eutrophication seldom does so.

As has already been shown, nutrient salts are deliberately added to British rivers in sewage effluents and as animal wastes, and accidentally when the excrements of farm animals are not properly disposed of. Modern arable farming also contributes. In some rivers, nitrogen levels have trebled between 1940 and 1970. This rise is, in some cases, partly caused by increased output of nitrogen in sewage effluent, but it is mostly caused by the loss of nitrates from arable land. This loss is a complex process. Between 1957 and now there has been a great increase in the amount of nitrogenous fertilizers used on cereals and grass. Total applications amounted to 381 000 tonnes of nitrogen in 1957, and nearly 800 000 tonnes in 1977. Only about half of this nitrogen has gone into the crops.

Nitrogen is lost from all soil regardless of whether or not it is fertilized with organic or inorganic manures, or is synthesized by nitrogen-fixing bacteria. The exact way in which losses occur is still uncertain. Some, but certainly not all, is leached out into the drainage water, particularly in the autumn and winter, and some is turned into atmospheric nitrogen by denitrifying bacteria. When farmyard manure was applied in bulk to bare soil in the autumn, the greater part of its nitrogen was lost during the winter. This had little effect on rivers, as the temperature was too low for massive algal growth and the salts passed harmlessly, but wastefully, to the sea. It is often assumed that salts in inorganic fertilizers will be more readily and rapidly lost than those in farmyard and other organic manures, and there is evidence that this may be the case, especially when the soil structure deteriorates, but some inorganic fertilizers, used at the right time, are efficiently converted into plant tissues. Thus, when grass is

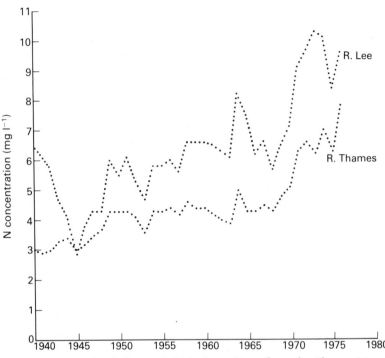

Fig. 3-3   Nitrate levels in two British rivers. (Data from the Thames Water
Authority.)

given a top dressing in early spring to stimulate its growth, the salts are
usually efficiently and almost totally taken up by the mass of fibrous roots
under the turf. However, this process can also give rise to a particularly
serious form of pollution. If the fertilizer is applied to saturated soil after
a wet winter and then there is heavy rain before the material is
incorporated in the soil, it may be washed off and give very high nitrogen
levels, as much as 30 ppm, in the rivers. Where these rivers are used as
water supplies this may be dangerous. Purification of river water removes
much of the pollution, but soluble nitrates usually remain. They are of
little danger to human adults, but young animals, including human
babies, have a different flora in the gut, and this may turn the nitrates to
nitrites which are poisonous, acting by combining with the haemoglobin
in the blood to form methaemoglobin which does not carry O_2 from the
lungs to the tissues. Fatal methaemoglobinanaemia in babies has
occurred in Europe and America, caused by drinking water containing
too much nitrate. In parts of East Anglia, wells in rural districts have been
found with dangerously high nitrate levels and one infant's death has
been attributed to this cause. Bottled water is imported for drinking in a
number of areas in Britain and the U.S.A.

To affect the aquatic vegetation, water must contain the various salts necessary for growth. The different components causing eutrophication come from various sources. Nitrates are leached mainly from agricultural land, but little phosphate, even when liberally applied in chemical manures, is lost as it is usually bound firmly in the soil. Phosphate levels have increased spectacularly since 1952, and these are believed to come mainly from detergents. There is a close correlation between the increased use of detergents and the rising phosphate levels in several British rivers. There is some argument as to whether phosphate or nitrate pollution has the most serious effects. They probably both make their characteristic contribution.

Eutrophication affects the vegetation of running water, but its greatest effects are seen when this water is impounded in a reservoir or where the river runs into a lake. The most obvious result is an algal bloom. Many species of algae are involved, sometimes unicellular forms (*Monodus* sp.) pullulate to turn the water into something like a pea soup, at other times the whole surface is covered with a mat of blanket weed (*Cladophora*). Algae give considerable trouble if the water is to be used for town supplies, as they choke the purifying plant. They also induce deoxygenation. This sometimes causes surprise, as they are green plants which produce O_2 during photosynthesis, so they might be expected to have the reverse effect. They do of course liberate O_2, but with blanket weed this is at the surface and much is lost, while the lower layers of the water are in shadow where plants cannot flourish. Finally the algae die and decompose, and this removes O_2, as does the decay of any other form of organic matter.

Algal blooms do not occur in pure water containing few salts, but they have been observed when nitrate–nitrogen levels are as low as 0.3 ppm and phosphate–phosphorus has been as low as 0.01 ppm. Such and even higher levels occur in many waters where blooms are infrequent, though algal growth is commonest where eutrophication is most intense. We do not always know what triggers off an algal bloom. In some cases grazing crustacea, including *Daphnia*, may keep the algae in check, and if these are killed by a spill of insecticide, the bloom takes place, uncontrolled. Sometimes, where nitrate is low and phosphate high, blue-green algae (e.g. *Anaboena* sp.), which can taint the water, are found in some abundance, for these plants are able to synthesize nitrates from atmospheric nitrogen. The whole question of the harmful effects of eutrophication and the conditions which encourage unwanted algal blooms, requires further study. Sometimes there is a balance, and algae, though growing rapidly, are kept at bay by being grazed by *Daphnia* and other crustacea. This cannot always be relied upon to continue, so every effort should be made to prevent nutrient levels rising suddenly as a result of pollution.

Some waters are rendered eutrophic when polluted by birds. Reservoirs and lakes are favourite roosting places for gulls, which sleep overnight in their thousands on the surface of the water. Unfortunately they also excrete into the water, increase its nutrient status, and precipitate algal blooms, with just as serious results as when man is the culprit. This has occurred for many years at Rostherne Mere in Cheshire, and more recently in the Norfolk Broads.

4 Thermal Pollution

Life can only exist over a comparatively narrow range of temperature. Warm-blooded animals, i.e. mammals and birds, have developed control mechanisms which maintain nearly constant body temperatures independent of those of their surroundings, but even they may be killed by extremes of cold and heat. The rest of the animal kingdom, and all plants, are profoundly affected by changes of temperature. Each organism has a cold death point, and a rather narrow range of temperature where life can proceed. As a rule, within this zone metabolism is slowest at the cold end and increases as it gets warmer. Above this zone are higher temperatures where exposure is unfavourable and, eventually, fatal. Laboratory experiments to determine the effects of high temperatures have only a limited ecological significance. Thus, a fish may live apparently normally in a fish tank, but may never be found in a river heated to the same temperature. The whole balance of animal and plant life in a habitat may be upset by a comparatively small change in temperature, so that an organism which appears to live normally in isolation may be eliminated by other species which find the new temperatures even more favourable.

If some activity increases the temperature so that man, plants or animals are harmed, this may clearly be considered to amount to 'thermal pollution'. In North America there are several widely reported cases of rivers being heated almost to boiling point so that they are completely lifeless. Other cases of water at 50 °C or above contain no fish and practically no invertebrates, though a few thermophyllic bacteria may flourish. Such examples are not widespread. Industry tries not to damage the environment unnecessarily. So far no really spectacular thermal pollution has occurred in Britain, but increasing industrialization is having its temperature effects and these need to be examined. The most notable are where water is used to cool machinery, particularly electrical power stations.

4.1 Dissolved oxygen levels

Biologists are particularly worried lest heating water may cause deoxygenation. If well-oxygenated water is warmed, the level of O_2 in it is reduced. This is because, as the temperature rises, the solubility of O_2 in water decreases. Thus, at 5 °C a litre of water can take up 9 ml^3 of O_2, while at 20 °C only 6 ml^3 will dissolve in the same volume. This imposes difficulties on an aquatic organism, for a rise of temperature from 5 to 20 °C is likely to increase its rate of metabolism, and therefore its rate of respiration (and its need for O_2) by a factor of about four. Of course, if the water which is warmed is, at the lower temperature, only two-thirds saturated, warming will not expel O_2 until a temperature above 25 °C is reached. Thus, heating polluted water may have less effect on the O_2 content than in a similar rise in temperature under purer conditions. The fauna of unpolluted water is also more easily destroyed by environmental change so, if it is used for cooling as industry spreads to rural areas, we must be on the look out for harmful effects.

In Britain, the electrical industry is the greatest user of cooling water. The scientists in the Central Electricity Generating Board research laboratories have made studies of the effects of power stations on the rivers which they use. Thus, at Drakelow on the River Trent some 850 000 m^3 of effluent from the cooling towers was returned to the river in summer, and 650 000 m^3 in winter. Less water was required in winter as it started off colder, and so could absorb more heat. The average temperature of the intake over 24 hours in summer was 18.2 °C and that of the outlet was 27.4 °C. In winter, the corresponding figures were 6.6 and 20.4 °C. Rises of this magnitude obviously had a substantial local effect, and temperature rises, to 22.4 °C in summer and 12.4 °C in winter, were measured 1.6 km downstream so a large area of the river was affected.

Contrary to common belief, this warming of the effluent has not caused deoxygenation, in fact the O_2 level in rivers has been improved. The Trent is a polluted river, with the O_2 level falling as low as 17% of saturation, and seldom rising above 50%. The water used in cooling is saturated or even, to some extent, supersaturated. As it is warmed, the amount of O_2 taken up is limited, but nevertheless at Drakelow between four million tonnes of O_2, in summer, and two million tonnes in winter, are added to the river every day.

4.2   Ecological effects

These changes in temperature and O_2 content of the Trent must obviously have ecological effects. Tubificid worms, able to live in the polluted water, appeared to breed more successfully below, rather than above, the power station, and the breeding season was extended, so that it reached its peak in October, when breeding in the cooler water above the station did not take place. In the less polluted River Severn near the Ironbridge power station, similar temperature changes occur. Their effects on the distribution and life histories of several larval stoneflies (Plecoptera) and mayflies (Ephemeroptera) have been investigated. Very little effect of the increased temperature was observed, though several species continued to grow only in the warmer waters during the winter. So far no significant changes in the balance of various species has been discovered.

In the U.S.A., where greater temperature changes have been produced, more obvious results have been observed. These have usually been, from the anthropocentric view, for the worse: salmon and trout have been replaced by coarse fish; the total protein production has been increased, but not the quality of the environment; and there has generally been a loss of species, and a reduction in diversity. Such changes in Britain may be expected if greater temperature changes are induced by an increase in the use of river waters for cooling. But so far, this would not seem to be a serious cause of what should strictly be called 'thermal pollution'.

Many nuclear power stations are on the coast, and are cooled by sea water. This does not seem to cause much damage. The water in the vicinity is certainly warmed and can be used for farming oysters and mussels, though such ventures have received limited support for fear that the molluscs may concentrate radioactive materials in the effluents. The main trouble has occurred within the cooling system, where young stages of mussels may pass through the grids and settle down inside the cooling tubes. They grow unusually fast in the warm water and interfere with the circulation.

Obviously it would be prudent to try to use all this surplus heat, perhaps for fish farming under controlled conditions, for district heating, or in the glasshouse industry. If this is done, further pollution will perhaps be avoided. But the main worry will continue to be that warming waters which are so far unpolluted may result in ecological degradation of a rather subtle and long-term variety. Gross thermal pollution, with the sudden death of fish from heat or deoxygenation, will be much easier to detect, and good long-term planning should prevent it from happening.

4.3   Global effects

To many people, thermal pollution signifies global changes of climate with adverse effects on man and the economy of the globe. Most of the energy produced by industry is eventually wasted as heat, and this must raise the temperature at least in the vicinity of its production. The question is whether such a rise is significant. We know that the climate of a large conurbation differs from that of the surrounding country; the average temperature may be as much as two degrees higher. This affects plant and animal growth, though results are not spectacular, perhaps because in the past smoke has reduced the light intensity and caused opposite effects.

At the moment, the heat liberated by all the power sources is an insignificant fraction of that which reaches earth from the sun. In the early 1970s, many authorities expected energy requirements in Britain to increase rapidly, even doubling every eight years – an exponential increase of a kind which could eventually reach, and even surpass, the heating capacity of the sun's radiation reached by this globe. Today, there is much less alarm from this cause. It is realized that exponential growth seldom continues for long and the curve soon flattens out. Also, since oil prices rose dramatically in 1973, energy use in many western countries, including Britain, has ceased to increase and, in some years, there has been a substantial fall. Recently, most forward estimates of energy needs have become more and more modest, as high costs have encouraged economy rather than waste. It is therefore unlikely that man-made heat will itself alter our climate, except locally. These effects hardly deserve the name of pollution.

The possibility that pollution will alter the climate of the whole globe by affecting the proportion of the sun's energy reaching it, requires more serious consideration. But pollution from high-flying aircraft does produce particles in the stratosphere and many pollutants including ammonium sulphate produced from SO_2 may induce cloud formation, and these could affect the 'albedo', or reflectivity, and so reduce the amount of radiation reaching the world's surface. Changes in atmospheric composition, particularly rises in levels of CO_2, could trap differing amounts of the energy which reaches the world. So pollution could cause the world to become cooler or hotter. We really do not know which. An international conference of the world's leading experts in February 1979 could come to no agreement, and did little more than draw attention to our ignorance. They could only agree on the importance of monitoring, on a world-wide scale climatic changes, and changes in the composition of the atmosphere at all levels, paying particular attention to the stratosphere. Many still considered it likely that natural fluctuations of temperature will be greater than those which man, with his present technology, can produce. But technology will develop, and there will be conscious efforts to 'improve' the world's climate. There is a risk that improvements in one region might be harmful in another.

5 Radiation

5.1 What is radiation?

There is no doubt that radiation can be harmful to life. It can also save life, when used for medical purposes. Dangerous radiation is released with lethal effects when a nuclear bomb is exploded, and many people fear that a nuclear power station might be equally dangerous. There is very great concern about radioactive waste from power stations, which may give off potentially lethal radiation for thousands of years. We are right to be concerned, though not to be hysterical.

Radiation is very complicated. It may consist of electromagnetic waves or bits of atoms moving at high speeds, which can damage living cells. Different radioactive substances produce various types of radiation, which have their own particular biological effects. It is only possible to give a brief outline of the subject here. Radioactive substances released into the environment may give off γ-rays, which affect the tissues outside the body, or they may produce β-rays, which only penetrate a few cell layers, and so are only effective when actually on, or in, the body. Finally, some produce α-particles, which deposit their large amount of energy over a very short path in the tissues. These α-particles are biologically the most damaging.

Radioactive substances do not go on emitting radiation for ever, they all 'run down' eventually. Some, including certain radioisotopes produced in a nuclear explosion, give out all their radiation in a fraction of a second. Some continue for thousands of years. The concept of the 'half life' is used in this connection. Thus, strontium-90 which gives off β-rays has a half life of 28 years. This means that at the end of 28 years the level of radiation given out falls to half, by 56 years to a quarter, after 112 years to an eighth, of the original level. Plutonium-239 has a half life of 24 000 years, so once released it is a very long-term hazard to the environment.

It is easy to be confused by the many units used to describe radiation. As we are concerned here with the biological effects of radiation on man and other organisms, the 'absorbed dose' of radiation seems to be the most relevant conception. Until recently the unit used here was the rad, which is still used by many writers. In the S.I. system the Gray (Gy) is now used, and conversion is easy as

$$1 \text{ Gy} = 100 \text{ rads}.$$

The unit of radioactivity used in the past was the Curie (Ci). This was a large unit, and activity often had to be measured, for instance from the fallout of a distant bomb, in picocuries $(pCi = Ci \times 10^{-12})$. The S.I. unit is the Becquerel (Bq) and

$$1 \text{ Bq} = 2.7 \times 10^{-11} \text{ Ci}$$

The biological effect of radiation obviously depends on its intensity. Thus a really high dose, say of 100 Gy, will be immediately fatal. A substantial dose, in the order of 2–6 Gy may cause immediate burns and other acute effects, and many people so exposed will die within six weeks. Low doses, or doses given to a small area of tissue may result in acute changes to skin, hair or bone marrow, from which recovery is usual.

However, doses lower than those which produce acute effects may result in delayed damage, of which the most important is the increased incidence of cancer, and the possibility of damage to the reproductive organs, giving genetic effects, only seen when irradiated parents produce abnormal offspring. It is generally accepted that there is no threshold below which no damage occurs, so any additional radiation may be harmful. Differences of opinion exist as to the danger from any particular level of radiation.

Radiation is something from which we cannot escape, as we are all exposed to 'background radiation' from natural sources: from cosmic rays (reaching the world from outer space), from most rocks (which have some small amount of natural radioactivity), and from deposits, particularly from the small amount of radioactive potassium naturally present, which are located within our tissues. The amount of radiation from rocks differs throughout the world, and even in different regions in Britain. The radiation in Aberdeen, a city built of granite, a rock with a

Table 1   Radiation exposure of the British population from all sources. (Data from reports of the National Radiological Protection Board, Harwell, England.) ($1 \mu Gy = 0.1 m$ rad.)

Source	Annual genetically significant dose (GSD)	
	μGy	%
NATURAL BACKGROUND		
approx. half from rocks, remainder equally from cosmic rays and from radioactive materials, particularly potassium in the body	870	83.8
MEDICAL IRRADIATION		
including X-rays and medical therapy	140	13.5
FALLOUT		
from nuclear testing	21	2.02
MISCELLANEOUS		
including occupational sources	7	0.67
DISPOSAL OF RADIOACTIVE WASTE		
(1974 figures)	0.1	0.01

comparatively high degree of radioactivity, is nearly twice that found in southern England. Table 1 shows that the natural background is responsible for 90% of the radiation reaching our bodies.

As man has evolved with this natural background, it is often assumed that it is harmless, and that the other sources, being so much lower, are of no danger. This point of view is not generally accepted by biologists. The natural background may be one of the dangerous natural hazards found on earth. Attempts are made to reduce all additions to the minimum.

5.2   Permissible levels of radiation

Permissible radiation levels are often related to the natural background levels. For food which might be contaminated from a controllable source, a figure of one thirtieth of the background is sometimes stipulated as the highest permissible dose to any consumer.

The International Commission in Radiological Protection (ICRP) makes recommendations to protect both workers in professions where radioactivity is a particular hazard, and the general public. They do not as yet set exact limits in all cases, but set forth the following general principles:

(a)   no practices (which produce increased radiation) shall be adopted unless their introduction produces a positive benefit;

(b)   all exposures shall be kept as low as reasonably achievable, economic and social factors being taken into account;

(c)   the dose to which an individual is exposed shall not exceed the limits recommended for the appropriate circumstances by the Commission.

ICRP's general guidelines are that radiation workers should not be exposed to more than 50 mGy (5 rads) a year, and that the maximum exposure for the general public should be one tenth of this (i.e. 5 mGy, 0.5 rads).

In fact there are two standards in industry and medicine: one for those involved in using radiation in their work, and a much lower one for the general public. The British Ministry of Labour in its 'Code of Practice' lists what are called 'designated persons' whose work exposes them to particular risk; they are generally allowed to experience a higher level of radiation than is generally acceptable, but it is usual to subject them, subsequently, to stringent safeguards so that a total dose of a dangerous level is not reached. Older men, with less risk of subsequent exposure, are often permitted to receive more radiation than their younger colleagues.

It should be stressed that although there is no firm proof of damage to individual men from these very low levels of radiation, good circumstantial evidence exists, for instance from the higher levels of cancer in radiologists than in the general population, and from the fact that, in the U.S.A., while the average age at death of the general population is 65.6 years, that of radiologists is only 60.5 years. The exact way that this exposure to radiation has affected the bodies of those concerned is still not fully understood.

Although those receiving specially high doses of radiation are clearly at risk, it is more difficult to assess the possible danger of existing levels of world pollution. It may sound very alarming when it is said that the level of some radioactive substance has increased, perhaps even by a factor of over a thousand, because of testing atomic weapons, but if even after this increase the absolute level of radiation produced is still only a tiny fraction of the background level then the risks may not be very great. Thus, it is true that radioactive materials have increased enormously as a result of testing nuclear weapons. This is illustrated in Fig. 5–1. Strontium-90 did not exist before it was produced in nuclear explosions. The graph shows that this nucleid, which is deposited on grass which is eaten by cows, appears in milk. The level in milk was low about 1960, but rose many-fold by 1964. Then, however, the Test Ban Treaty took effect, and the levels fell right down by 1970 and have remained at that level.

Fig. 5–1   Country-wide average ratio of strontium-90 to calcium in milk in Britain. (Data from Annual Report, A.R.C. Letcombe Laboratory, England.)

Although the graph looks alarming, it measures very small amounts of radiation, as is indicated in Table 1. Even at its worst the extra radiation was a tiny fraction of the background level. However, had above-ground nuclear bomb tests continued, levels could have continued to rise to those which were really dangerous.

5.3   Biological effects of radiation

There are, however, additional complications. The danger from any radioactive substances depends on where they are located. Strontium-90 is often spoken of, somewhat emotionally, as a 'bone seeker'. This is intended to indicate that it may be concentrated by man from his food or drink, and stored in his bones. There is not any special concentration of radioactive substances. Bones normally contain much calcium and a substantial fraction of strontium. Man and animals naturally concentrate these elements into their bones and, if radioactive strontium is present, it is also concentrated together with the normal element. The danger from strontium-90 is particularly great because it is concentrated adjacent to the bone marrow. The marrow is active in blood production, and radiation may cause a type of cancer which manifests itself in the fatal disease leukaemia. When weapon testing began, there was a slight world-wide increase in the incidence of leukaemia, and some scientists correlate this with radiation and in particular with this increase in strontium-90. The risk to any one individual is still so small as to be almost unmeasurable, but nevertheless several thousand cases may well have developed among the whole of the world's population. This, and the risks from other isotopes, is sufficient to make the complete cessation of weapon testing desirable. Since above-ground tests have been restricted, world radiation levels have in fact decreased greatly, as is shown in Fig. 5–1, but could easily increase again if the restrictive agreements broke down or if more countries evaded them.

If we survive nuclear testing and the almost certain extinction which would result from nuclear war, we will still be subject to man-made radiation from medical and industrial sources. We are becoming increasingly stringent in our control of X-rays and of radiation used in medicine, and have reached a situation where any further reduction would mean abandoning many useful techniques. It is generally agreed that less damage is likely to the health of mankind from the present levels of radiation than would be inevitably result from abandoning these techniques for medical diagnosis and treatment.

5.4   Nuclear power

The whole question of the use of nuclear power is giving rise to controversy throughout the world. Many of the points were raised in the prolonged Windscale Inquiry into the nuclear power industry in Britain in 1978. Coal, oil and natural gas are being exhausted, and larger amounts of energy are needed if developed countries are to progress and developing countries are to become even partly developed. Sun, wind and the tides will undoubtedly make a greater contribution, but with present technology they cannot fill the gap when oil is mostly used up early in the twenty-first century. The protagonists of nuclear power believe they have the answer, particularly if Fast Breeder Reactors, which economize in the use of scarce uranium fuel, are built. On the other hand, most of those who call themselves 'environmentalists' are totally opposed to nuclear power, on the ground that it is too dangerous. The danger comes from radiation, both from the operation of the power stations and from the radioactive materials they produce.

It is generally agreed that existing nuclear power stations have an excellent record, with far fewer deaths than occur when coal is mined or oil is drilled. Very little radiation has escaped into the environment, as indicated in the table. Yet there are many problems in processing radioactive waste, and of disposing of the products. Today in the U.S.A. some 150 m³ of the waste produced annually contain over 1000 MCi of radioactivity. A megacurie is the activity of a million grams of radium; only some 10 000 g of radium itself have in fact been isolated for all medical uses in hospitals in all countries. The size of the problem may be indicated when it is realized that the levels of radioactivity in vegetation, etc., polluted by nuclear fallout are measured in picocuries (or micromicrocuries), and that one megacurie is 10^{18} pCi. By the year 2000 the annual level of radioactive waste in the U.S.A. may reach nearly 1 000 000 MCi, and energy requirements will no doubt increase after that date, so disposal will become an even more serious problem. Incidentally, for the purposes of this argument it will be noted that I have continued to use the old unit of activity, the curie (or the megacurie), as the S.I. unit – the Becquerel – is so small that, in this connection, a great many 'noughts' would have been needed to describe the situation. (For conversion formula see p. 40.)

There is strong opposition to erecting any nuclear power stations, partly because they themselves may be dangerous, partly because of radioactive waste, but mainly because they produce materials which could be used in nuclear weapons, perhaps by terrorists. It is nuclear war, which could exterminate mankind, which is the great danger. It can be argued that most countries could build nuclear bombs whether or not the nuclear power industry expands, and that mankind might as well derive the maximum benefit from this source of energy, remembering that if it works as planned it causes far less pollution than does the use of fossil fuels.

However, the larger the number of power stations, the greater the risk of an accident, with the release of a dangerous amount of radiation. It seems likely that with an enormous number of these installations in all countries, it will be difficult to ensure that all observe the type of precaution at present in force in the few existing stations, particularly with regard to the disposal of radioactive waste. Thus although nuclear power has, so far, done little harm, in the long run it might well be the most dangerous form of man-made pollution.

5.5 Mutations

There is much concern lest radiation should cause harmful mutations, particularly in man. However, we have little evidence of the results of low levels of radiation in actual practice. There is no doubt that moderate intensities of radiation can cause mutations. This has been demonstrated in the laboratory with the fruit fly *Drosophila*, with mice, and with various crop plants. The mutations produced are almost all harmful, producing deformed organisms. Lethal genes, causing the embryo to die, have also been produced. But the level of radiation has usually been higher than would be tolerated in the general environment for health reasons. In theory, much lower levels could be effective, but this is difficult to prove by animal experiments as such huge numbers of animals would be required. Thus, we have little hard evidence of the mutagenic effects of man-made radiation outside the laboratory. The background radiation may have been an important factor in producing the mutations on which the whole process of evolution depends, but this again is a speculation. The reason for measuring radiation as the genetically significant dose (GSD), as in Table 1, is to relate radiation to its effects on the gonads and so, eventually, to possible mutations.

One experiment suggests that radiation-induced mutations may not be as frequent as is often imagined. In the Brookhaven Forest in the U.S.A. a fascinating experiment has been in progress for the last 20 years. A massive source of radiation switched on for some 20 hours of every day has killed all life in its vicinity. Further away, more selective damage has been done, and valuable results on the susceptibility of different plants and animals have been obtained. Further still from the source, the radiation, though far above the levels tolerated for human exposure, causes no obvious damage. Nor has it apparently produced any obvious mutations in the plants and animals subject to these different levels of exposure. This may indicate that the risks of producing mutations may have been overestimated. However, as quite low levels of radiation can obviously cause damage to man and to other organisms as is revealed by epidemiological studies in a large population, even where no obvious damage can be observed in an individual, it is obviously right to try to keep radiation as low as is practicable.

6 Pollution of the Sea

The oceans cover 56×10^6 sq km and contain 1420×10^{15} m^3 of water. This is an awful lot of water! It corresponds to nearly 5×10^8 m^3 for every inhabitant of the earth; it is not surprising, therefore, that we often assume that the sea can absorb unharmed all the wastes we like to discharge into it. Unfortunately, however, pollution of at least some of the seas does occur, largely because noxious substances do not become equally mixed but remain concentrated in limited areas, or may even be reconcentrated by living organisms after apparently safe dilutions have been achieved.

6.1 Oil pollution

Oil is the most obvious pollutant of the ocean. It has caused damage in the open ocean and in estuaries, and even where spills have occurred in rivers well inland. The dangers from this source received great publicity after the tanker, the *Torrey Canyon*, went aground on rocks off the coast of Cornwall on 18 March 1967, and when more than half its cargo of nearly 120 000 tonnes of oil escaped. Since then, we have had even worse spills, such as the *Amoco Cadiz* in 1978, when miles of beaches in Northern France suffered. There have been serious leaks from the oil rigs in the North Sea, and from terminals such as that at Sullom Voe in the Shetland Isles where the oil is brought ashore. There have been many other serious, if smaller, incidents when tankers have been wrecked or damaged, when ships have deliberately discharged waste oil, and when tanks have been washed out prior to reloading.

The oil industry is so huge that it is hardly surprising that accidents occur. Tankers now carry annually some 8×10^8 tonnes of crude oil around the world, and it is said that several million tonnes of waste oil are today floating on the oceans, often harmlessly, but still a potential danger if washed ashore or into a region where birds are diving to catch fish. Until recently it was policy to wash tanks out at sea, adding several million more tonnes to the ocean's pollution. This has been decreased as the 'load on top' system has been adopted by most of the leading oil companies. In this, the oily waste at the bottom of all tanks after washing is collected in one tank. Here the oil floats to the top of the water which can then be safely pumped out from underneath. The oil is retained and the next load is added to the tank without further loss.

Oil is a serious pollutant in its own right. The effect on sea birds is well known. Their feathers become clogged, and they cannot fly. In attempts to clear the plumage, the birds preen themselves, and in doing so imbibe sufficient oil to poison themselves. The oil interferes with the insulation provided by the feathers, and the birds can die of cold or their susceptibility to pneumonia increases. The *Torrey Canyon* incident alone may have killed 100 000 birds of different species; guillemots were the most severely affected.

The results of oil pollution on birds are distressing, and there is no doubt that they are harmful to many individuals. The effects on the populations of the different species are more difficult to assess. Guillemots and other auks are now less common on the coasts of much of southern Britain than they were 30 years ago, and this may be the result of oil pollution. However, it must be acknowledged that no species is in danger solely as the result of oil, and the ability of many to increase after there have been numerous deaths is remarkable. So, though oil pollution does seriously affect many birds, its result on a world scale may not be as serious as is sometimes imagined.

Oil affects many other marine organisms. Where rocks are thickly covered with oil, seaweed dies, and many molluscs and crustaceans are also killed. But the remarkable thing is that the recovery, at least from a moderate amount of oil, is remarkably rapid and complete. Bacteria 'live on' the oil and break it down. Within a year of quite heavy oiling, rocks and beaches have been almost completely restored, and their flora and fauna have regenerated. The most serious damage has been done by the attempts to combat the oil pollution itself.

The great economic damage done by oil is when it lands on the beaches of seaside resorts. For years tarry patches have soiled the skin and the scanty bathing attire of holiday makers and dirtied the feet of paddling children. The problem is an unpleasant one even when the oil is present in small amounts. A massive spill can cover the whole surface of the sand and drive all holiday makers away. Various detergents have been used to emulsify and disperse the oil. They have been reasonably successful in this task but unfortunately they have proved to be much more toxic than the oil itself to many marine organisms.

The toxicity of detergents was realized before the *Torrey Canyon* incident, for it had been observed that where part of an oiled rock had been treated, recovery of the flora and fauna was slower than where the oil was left untouched. However, the possible economic damage of the *Torrey Canyon* spill was such that some 3×10^8 dm^3 of various detergents were used to clear and protect Cornish beaches, notwithstanding the reservations of the biologists. As the substances used (nonyl-phenol polyoxyethylene condensates) are toxic to crustaceans at levels well below 1 ppm, it is not surprising that considerable damage was done. The unexpected thing is that it was not much more serious. Although many species of molluscs, crustacea and fish suffered heavy casualties, and large areas of rock were denuded of their seaweeds, yet recovery was remarkably rapid and within two years was virtually complete. No permanent losses of any species of animal or plant were reported.

The oil from the *Torrey Canyon* would have done more damage in British waters had not the wind changed before all the oil had been blown ashore and diverted it to the coasts of France. Here considerable local damage was done to fish and oysters, but little detergent was used. Instead the water was covered with sawdust and chalk. Oil which was sunk in this way did some damage to the organisms on the bottom, but again it was broken down fairly quickly. More oil was washed ashore and trapped in the salt marshes. Here again local and transitory damage occurred, though examination in 1968 suggested that some species were more vigorous, their substrate having been 'fertilized' by the oil.

These studies should not be allowed to suggest that oil pollution is unimportant. Had the whole of the *Torrey Canyon*'s load gone ashore rapidly in Cornwall, as it might have done with slightly different weather conditions, the results would have undoubtedly been more serious and longlasting. The continued seriousness of the problem was highlighted when in 1978 the oil from the *Amoco Cadiz* did real damage in France, as did that from the *Eleni V.* in eastern England. It is now generally agreed that detergents should only be used on beaches frequented by holidaymakers, as natural processes remove oil fairly rapidly unless the concentration is too high.

6.2   Sewage and poisonous wastes

A large amount of raw sewage is discharged into the sea. Where long outfall pipes are used, so that there is little risk of fouling beaches, this system is reasonably unobjectionable. The organic matter breaks down and the nutrients are recycled. However, the discharge is often too near to the land, and the shore may be fouled. This is unaesthetic, and there must be some risk to public health, but this is less than is often imagined. A thorough investigation in towns in Britain where such pollution of the beaches was common showed that, even among those who swam among the sewage, there was little evidence of infection from pathogenic bacteria or viruses.

Other organic wastes are dumped into the sea. Freezing plants in Lincolnshire discharge considerable quantities of vegetable wastes, but the results are comparatively small. Reasonable mixing seems to occur, though the local enrichment may promote the growth of mussels and some seaweeds.

Today the main concern is that the ocean may be being polluted by poisons which do not break down as rapidly as does sewage or other organic materials. There are reports of discharges of heavy metals, including mercury and lead, of industrial chemicals such as polychlor biphenyls, and of organochlorine insecticides. These last are dealt with in Chapter 7.

There is insufficient control of the dumping of poisonous waste materials outside territorial waters, and international agreements are not always effective. Considerable industry exists to get rid of embarrassing wastes. In many cases this is done in a responsible manner, fishing grounds are avoided and durable containers likely to prevent leakage, at least in the immediate future, are used. Other dumping, however, is less well-controlled, and local pollution has occurred.

Any substance completely mixed throughout the oceans would be diluted to a harmless level. Even if a million tonnes were discharged, the level throughout the oceans would be less than one part in a million million, and no pollutant is known that is toxic at that level, or that could be concentrated by any known organism from such a dilution.

Where mercury used in wood pulp manufacture has been discharged into estuaries (e.g. in Japan), levels have been high enough to damage molluscs and fish, which have concentrated the mercury to a level where men eating these animals have been fatally poisoned. There is clearly a considerable risk from these discharges into estuaries or to enclosed waters like the Baltic. Local damage to fisheries in shallow areas like the North Sea may occur before sufficient dilution has occurred. Some years ago it was suggested that tuna fish were seriously contaminated by mercury discharged into the ocean by man. It now appears that the mercury occurred naturally in the ocean, and that man-made additions were negligible in comparison with the amount of the metal already present. The tuna were able to concentrate the naturally occurring metal. Though 'natural', this could be as toxic as if it had been introduced by man. Fortunately the mercury levels in tuna, whatever their origin, do not seem high enough to endanger the eater.

It has been suggested that lead also may be dangerous, if concentrations near to the coast are high enough. This could be derived from the tetraethyl lead in petrol. Local pollution may occur in estuaries and affect isolated animal populations. However, there is little evidence that the alarmist stories that the whole fauna and flora of the ocean may be so seriously polluted by metals and organic poisons that it is being destroyed are true. Nevertheless, it would be prudent for international dumping and monitoring schemes to be enforced, to measure changing levels of persistent pollutants in different parts of the ocean, and to investigate changes in animal and plant populations.

7 Pesticides

7.1 The pesticide problem

The pollutants so far considered have been substances produced by man or his industries, which have caused unintentional damage to the environment. Their control has been achieved by taking greater care over their disposal and of their dispersal. Pesticides, on the other hand, are poisonous substances deliberately disseminated in order to exploit their toxic properties; they become pollutants when they reach the wrong targets.

Pests are organisms which man considers to be harmful, and pesticides are the substances used to control them. The main classes of pesticides are: herbicides, used to kill weeds; fungicides, which control pathogenic fungi; nematocides, controlling eelworms; and insecticides, developed for use against insects.

A pesticide is normally used against a particular organism. Ideally it should poison it, but be otherwise harmless. Although there are many chemicals with a remarkable degree of selectivity, so that the chosen pest is destroyed by a lower exposure than that required to damage other plants or animals, nevertheless complete selectivity is virtually impossible. This means that there is always the risk that pesticides will cause damage to man or to other, non-target, organisms.

Some pesticides are acutely poisonous but unstable substances. They can cause serious damage over a restricted area, but not long-term pollution. Other pesticides may be less acutely poisonous, but they may be much more persistent, and they may thus continue to have ecological effects for very long periods. They may be transported over long distances, and cause damage far from the original site of application. Also, a persistent poison may reappear where least expected. Although it may have been diluted down to a harmless level, it may be reconcentrated in some biological system up to levels where serious damage can be demonstrated once more.

It was in the 1950s that concern about the harmful ecological effects of pesticides became widespread. In Britain in spring, many seed-eating birds were found dead in the fields, and predatory birds feeding on them were wiped out in parts of the country. Although scientists were aware of these problems, and took effective action in some cases, public awareness was stimulated most by Rachel Carson's book *Silent Spring* published in 1963. For the next ten years pesticides became a major source of worry, some people fearing that the whole of the life-support system of the earth might be destroyed. It was feared that levels of very persistent insecticides, like DDT, would rise in the oceans and exterminate the phytoplankton and render the waters sterile.

The effects of public concern were both good and bad. Many of the most toxic and long-lasting pesticides are no longer used on a wide scale. Levels of DDT and similar chemicals have all fallen in the oceans and in the tissues of human beings in most parts of the world. New chemicals of very low toxicity to mammals and birds have been introduced. In Britain, all new chemicals have to be approved by scientists serving on committees of the Pesticides Safety Precautions Scheme. Many people are misled by the growth in the number of pesticides avilable for use. They see that in 1944 there were only 63 approved pesticidal products, but by 1976 this number had increased to 819. What the figures do not reveal is that the small numbers of pesticides available in the early years included very toxic preparations containing, for instance, arsenic. The most deadly and dangerous substances have either been banned, or their use has been restricted, and since the scheme has been fully operative there have been no serious incidents involving wildlife from newly approved substances.

The bad effect of public concern has been that DDT in particular has been banned or discouraged where it could have been valuable, particularly in saving human life. This substance, as is shown later in this chapter, is very persistent, and has caused serious damage to wildlife. But it is relatively non-toxic to man, and though hundreds of thousands of tonnes have been used to control malaria, typhus and other diseases, no one has ever died from poisoning except when they have eaten DDT thinking it was flour. Many deaths from disease which could have been prevented have occurred. Also, much more toxic insecticides have been used, in the belief that they were 'safer'. Fortunately, these dangerous chemicals are less used today, and disease control is being tackled more scientifically, but it is clear that the effects of uninformed public concern can be dangerous.

Herbicides are much more widely used than the other classes of pesticides. In Britain in 1976 about £50 000 000 were spent on herbicides, compared with about £8 000 000 on fungicides and the same sum on insecticides. Nematocides, suitable for use on a field scale only, became available in the late 1970s, and so far no harmful side effects on non-target organisms have been reported.

In the following sections different types of pesticides are dealt with systematically. As already suggested, some insecticides, like the organochlorines, are less important than they were in the 1950s, '60s and early '70s, but they are still dealt with at some length as they illustrate the principles of environmental contamination.

7.2   Herbicides

Herbicides seldom cause serious environmental pollution, though they are used more widely and in greater quantities than any other pesticides. In Britain the 'hormone herbicides', the phenoxyacetic acids, are the main means of controlling weeds in cereal crops. MCPA (4-

chloro-2-methyl-phenoxyacetic acid) is the most popular substance. Properly used on a cereal crop, the broadleaved weeds are eliminated with very little in the way of side effects. MCPA is not a direct 'killer'; it disorganizes the growth of the weeds which are then removed by drought and competition. The cereal seedlings are usually unharmed, though overdosing or spraying at the wrong stage of growth causes some upset. The final results are similar to the laborious weeding which preceded the use of these chemicals. There are some risks when a spray is applied carelessly or in windy weather, when the MCPA drifts onto other areas. Little in the way of 'run off' generally occurs, though spraying into ditches or ponds may kill aquatic and emergent plants. The effects on the soil flora and fauna are slight; the most noticeable is that bacteria, which take part in the breakdown of the herbicide, increase in plots which are repeatedly sprayed. Otherwise, worms and mites in the soil seem to be unaffected. The chemical is not persistent and so long-term contamination does not occur.

Another related herbicide, 2,4,5-T (2,4,5-trichlorophenoxyacetic acid), has enjoyed some notoriety because of its use as a defoliant in Viet Nam. Here the herbicide has been sprayed, mainly from the air, at a far greater concentration than has ever been used in agricultural practice. Also, the samples used have been found often to contain a high proportion of a very toxic impurity, dioxin (2,3,7,8,-tetrachloro-dibenzo-para-dioxin) which, at least in the laboratory, is teratogenic (i.e. if applied to pregnant animals, the young may be born deformed). It seems that, in Viet Nam, long-term ecological damage has been done, with the destruction of large areas of mangrove swamps and the death of forest trees. It is suggested that humans have also been harmed, and babies born deformed. However, with careful use of preparations which are not heavily contaminated with dioxin 2,4,5,-T can be a useful chemical particularly for the control of scrub. It would seem best not to use it as an aerial spray, where control of the quantity used and of the target is difficult, but otherwise it need not be thought of as a dangerous pollutant.

Other widely used herbicides are the persistent and long-acting substituted ureas (e.g. monuron) and substituted triazines (e.g. simazine), and the quick-acting but non-persistent bipyridylium compounds (e.g. diquat and paraquat).

Simazine is used in high dosage as 'total' weedkiller on railway tracks and similar places. It acts slowly, direct spraying on plants often seeming harmless. It usually acts through the soil, from which it is not easily leached. Simazine is also used to prevent weed growth among crops, for example maize and asparagus, which are not susceptible to it. Very little accidental damage has ever been reported when these chemicals have been used carefully, and they are unlikely to act as pollutants.

Paraquat gives results rather similar to a flame gun, killing the above-ground parts of most plants. There is some translocation through the plant, and grasses are rather susceptible, though deep-rooted thistles and convolvulus seem to recover with increased vigour. Paraquat is normally immobilized almost immediately by absorption onto the soil particles, from which it is not released, and is then broken down by bacterial decomposition. If therefore it is carefully applied to its target, there is practically no risk of pollution outside that area. One great advantage of paraquat, which is used to control weeds when pasture is resown, is that the soil fauna, particularly the earthworms, survive its application far better than they survive normal ploughing and cultivation. However, this chemical has disadvantages and dangers when wrongly used. First, it is very poisonous to man if accidentally drunk in undiluted form, in fact it is the only pesticide in use in Britain today which has caused deaths in recent years. Its action is particularly unpleasant, causing asphyxiation by damaging the lungs which are choked by the proliferation of the epithelial tissues. Eye irritation is also caused by exposure to low volume (and therefore concentrated) spray. Secondly, the chemical is much less rapidly immobilized in wet vegetation, and when sprayed onto weedy stubble after rain wild animals, particular hares (which are more susceptible than rabbits or cattle) have often been killed. But with proper and careful use, paraquat should not be an environmental pollutant.

Two herbicides which have recently been introduced have caused some concern. The first is asulam, first used to control dock plants. It has also been found to be effective against bracken. It is not an environmental pollutant, and is relatively non-toxic to man and other vertebrates, but it may be used to alter completely the appearance of much hill country where the bracken has an aesthetic appeal.

The second recently-introduced herbicide is glyphosate, which was only approved for use in Britain in 1975. It is a very powerful herbicide, being translocated within plants, even killing the underground parts of couchgrass and bindweed. Here again, it is the efficiency of the chemical which causes concern. It, like asulam, is of very low toxicity to vertebrates, and it does not remain active in the environment being firmly bound (and thus ceasing to be active) as soon as it reaches the soil.

7·3   Fungicides

Considerable crop losses may be caused by parasitic fungi diseases. These may be controlled by many different chemicals. Several preparations containing copper have been used for various blights and mildews which are caused by fungi. When applied in high concentrations for many years, as in apple orchards, the soil may become heavily contaminated with copper so that the soil fauna is affected, and worms (for instance) are eliminated, but little damage seems to be done to the mature trees and no serious losses of the copper to contaminate surrounding areas have been reported. Various organic fungicides (e.g. captan) are very effective, particularly in horticulture, but as their total usage is not great as compared with substances like MCPA, and as they are generally rather expensive (and therefore used with some care), no serious cases of pollution have apparently occurred.

Various compounds using mercury are used as fungicides. Calomel (HgCl), which itself is comparatively non-toxic so that it was formerly a popular purgative prescribed for human consumption, has been used in horticulture in fairly large amounts. Organomercury compounds are now used as cereal seed dressings to such an extent that, in Britain, seed corn is normally so dressed unless the customer stipulates otherwise. These seed dressings are used prophylactically, i.e. they are used to prevent common fungal infections, whether or not they are likely to cause damage in a specific field.

Mercury can be an important environmental pollutant. It is stored and accumulated by animals and may be transmitted and concentrated in food chains. There have been many cases where fish and molluscs have accumulated quite high levels of mercury, and even deaths have occurred in men eating the fish and molluscs. However, the serious cases of contamination seem to have arisen from the much more massive industrial usage (e.g. wood-pulp processing), than from the comparatively small amounts used in agriculture (see §6.2, p. 49).

Seed dressings are unlikely to contaminate the soil seriously. Only about 100 g of active ingredient are applied to a hectare, and the total addition of mercury, per square metre of soil, is in the region of 1 mg. This is often negligible in comparison with the natural level of mercury in the soil.

Some contamination of wild birds occurs when dressed seed is eaten. In Britain, the main dressings have been in the form of phenyl mercury compounds, which are not very toxic to vertebrates, and feeding experiments with pheasants suggest that they would never be likely to take up a dangerous dose. In some European countries methyl mercury seed dressings have been more widely used; these are much more toxic, and have been shown to kill birds accidentally. Even in Britain, contamination is not negligible, particularly as concentration in predators eating the seed-eating birds does occur, and may approach the level where physiological effects can be detected.

Different mercury compounds differ greatly in toxicity. One particular danger of this metal is that even where one of the least dangerous compounds, say calomel, is liberated into the environment, this may be transformed by bacterial action to the very toxic methyl form. This could occur within the gut of a ruminant, where methylization is known to occur, or in the bottom of a lake or in a swamp. For this reason, every effort should be made to reduce the levels of mercury. However, this applies more to industry than to agriculture, and it should be remembered that the vast bulk of mercury, in soils or in the sea, occurs as a result of the natural weathering of mercury-containing rocks. Although it occurs naturally, such mercury is as dangerous as that accidentally distributed by human actions.

In the 1970s systemic fungicides began to be used extensively. These do not present pollution problems, nor are they highly toxic. There are moves to remove mercurial fungicides from the list of approved chemicals, but it may be some time before these are finally implemented. Nevertheless, the trend is for modern chemicals to be less likely to cause ecological damage than those used before 1970.

7.4   Insecticides

Many poisons have been used to kill insect pests. Pests of fruit trees have been gassed with cyanide and sprayed with arsenic or with dinitroorthocresol (DNOC). Nicotine was widely used, particularly in glasshouses, where it is as poisonous to insects as to cigarette smokers. Naturally occurring substances like the fish poison, derris, have been found to be as effective against human lice and the raspberry beetle. Any of these substances could be a pollutant and have harmful side effects, but in practice serious environmental pollution by insecticides is a post-war (1945) problem, and dates from the widespread use of new chemicals which have been synthesized by man. With few exceptions, these synthetic insecticides have been much less acutely poisonous to man and other vertebrates than some of the substances which they replaced, but the great persistence of some chemicals has produced new problems.

The insecticides in wide use today fall into two main groups, the organophosphorus and the organochlorine compounds. Some of the organophosphorus insecticides are the exceptions to the rule that synthetic insecticides are relatively non-poisonous to man. Parathion (diethyl p-nitrophenyl phosphorothionate) and TEPP (tetraethylpyro-phosphate) are very poisonous, they may only be applied by operators using protective clothing, and they have caused thousands of accidental deaths and numerous others when taken deliberately as suicide drugs. However, though they have caused environmental damage when carelessly applied, they are unstable and soon break down to relatively harmless products and so they cannot cause long-lasting pollution.

Today, particularly in Britain, the very toxic organophosphorus insecticides are little used, as they have been largely replaced by safer substances. These include malathion (o,o-dimethyl S-1,2-di (ethoxycarbonyl) ethyl phosphorodithioate), which is less than a thousandth as toxic as parathion, and various systemic insecticides such as schradan (bis(tetramethyl-phosphorodiamidic) anhydride) and demeton methyl or 'metasystox' (dimethyl S-2-ethylthioethyl phos-phorothiolate), which are translocated through the plant and so are valuable in controlling aphids and other sap-feeding pests without directly endangering beneficial insects which do not feed on the plants. Carbamates, chemically distinct, but similar in action and persistence to organophosphates, are increasingly being used. As a whole they also do not cause lasting pollution. Today chemists are looking for substances which have a longer period of activity, and some of these may be more acutely toxic than, for instance, malathion, but we are now aware of environmental dangers and bear them in mind when assessing new organophosphorus or carbamate insecticides.

7.5   Organochlorine compounds

It is the stable and persistent organochlorine insecticides which have caused, and which are still causing, environmental contamination which at least sometimes is of sufficient magnitude to deserve the description of pollution. The insecticides most concerned are DDT (1,1-bis(*p*-chlorophenyl)-2,2,2-trichloroethane), or lindane (1,2,3,4,5,6-hexachlorocyclohexane) and members of the cyclodiene group, including dieldrin or HEOD (1,2,3,4,10,10-hexachloro-6,7-epoxy-1,4,4a,5,6,7,8,8a-octahydro-*exo*-1,4-*endo*-5,8-dimethanonaphthalene), aldrin or HHDN (1,2,3,4,10,10-hexachloro-1,4,4a,5,8,8a-hexahydro-*exo*-1,4-*endo*-5,8-dimethanonaphthalene), and heptachlor (1,4,5,6,7,10,10-heptachloro-4,7,8,9-tetrahydro-4,7-methyleneindene).

The insecticidal properties of DDT were recognized in 1939, and during the 1939–45 war malarial mosquitoes and typhus-bearing lice were controlled by it. After the war, production was greatly increased, and DDT was widely used in controlling both medical and agricultural pests. It was found to be remarkably non-toxic to man, and there is no case of a human death from the proper use of this insecticide. There is no doubt that DDT has saved millions of lives from malaria and other insect-borne diseases in tropical countries, and that it has saved many from starvation by increasing food production by destroying crop pests. It was only when it began to be over-used, and when dieldrin, which is rather more toxic to vertebrates, was found to be killing many forms of wildlife, that concern about these chemicals began to be felt.

There is little doubt that DDT can now be detected almost everywhere in the world. It occurs in air and in rainwater, at levels of a few parts in 10^{11} and at rather higher amounts in the fat of birds and fish even in such unexpected places as the Antarctic. These low levels are not in themselves apparently having any biological effects, so they cannot be considered as pollution in the ordinary sense, though they are a cause for concern as we do not understand how they arose, or what is the route by which the insecticides got into the environment. The main worry, however, is lest concentration should occur, raising the levels to those which are indeed harmful. This occurs because organochlorine pesticides are very sparingly soluble in water, but extremely soluble in fat, in which they may often be stored. The stored insecticide may do little harm, but when the fat is metabolized and the chemical is released into the bloodstream, its effects may be dramatic.

There have undoubtedly been many cases of harmful pollution by organochlorines, with the death of non-target organisms. Thus in the

U.S.A. massive amounts equal to over 30 kg^{-1} of DDT were used against the beetle carrying the spores of the Dutch Elm Disease; this almost eliminated the American robin mainly because it ate worms which had accumulated the insecticide. In Britain there is little direct evidence of unintentional deaths to birds or mammals from acute DDT poisoning, but when dieldrin was used as a seed dressing to protect young plants from attack from the wheat bulb fly larva, this seed was eaten by pigeons, pheasants and other birds, which died in large numbers. When the corpses were eaten by predatory birds and by foxes, these suffered fatal secondary poisoning. Some of the predators died from eating only one or two heavily contaminated prey, others concentrated the substances of many sublethal doses over a period from many apparently normal prey.

DDT is not particularly poisonous to man, for the acute toxicity is similar to aspirin. In fact, as mentioned earlier in this chapter, no one has ever died from DDT poisoning when the chemical was properly used, and few humans have been seriously ill, though some workers in chemical factories have suffered temporary indisposition from carelessly handling the substance. Nevertheless, almost everybody in the world has a measurable amount of DDT and its breakdown products, including DDE, in their body fat. There is no evidence that this is harmful, but it is satisfactory to know that body levels are falling. The amounts of DDT in food, from which our body residues are derived, are also falling, as are those of other organochlorine and organophosphorus insecticides. Table 2, giving the daily dietary intake of several pesticides, shows how our intake is falling: it also shows how far our intake is well below that thought dangerous by medical opinion.

Table 2   Figures for the acceptable daily intake, and the daily dietary intake, of some insecticides in Britain between 1966 and 1977. (Figures from the Report of the Government Chemist.)

Pesticide	Acceptable daily intake µg day^{-1}	Dietary intake		
		1966–7	1970–1	1974–5	1975–7
HCH	700	6.6	5.5	4.4	3.9
Dieldrin	7	6.6	2.1	1.9	2.3
DDT + TDE + DDE	350	44	15	12	5
Malathion	1400	11	53	7	7

There is no evidence that these levels in man – up to 25 ppm in their fat, much lower in the tissues – are having any physiological effects. Volunteers who over many months have deliberately ingested large amounts of DDT, and factory workers who have done so unintentionally over periods of years, have had fat levels of 100 ppm or more, again without showing any clinical symptoms such as have been observed where DDT has been mistaken for flour and consumed on that scale.

Levels of organochlorine residues in wildlife are often higher than in man. It is sometimes difficult to evaluate these residues, as many different chemicals are involved. This is partly because chemicals are transformed within animal tissues. Thus DDT is largely turned into DDE, in a process of detoxification. In fact DDE is still insecticidal and poisonous to vertebrates, though less than DDT, except that certain processes in the breeding animal may be more rather than less severely affected.

There is good circumstantial evidence that the sublethal amounts of DDT (and DDE), as well as those of dieldrin, have had their effects on wildlife. Thus it was shown that, in the late 1950s, there was a dramatic reduction in the thickness of eggshells of those predatory birds which accumulated the greatest body burden. Peregrines became rare, and almost ceased to breed in southern Britain, in the areas where intensive cereal growing was practised (Figs 7–1 and 7–2). In Scotland, golden eagles became sufficiently contaminated to affect breeding, though the population of adults (which may live for 30 years or more) has apparently not been significantly reduced. It is also encouraging to be able to report that, since the use of dieldrin as a seed dressing and a sheep dip has been curtailed, a significant recovery in the breeding success of both hawks and eagles has occurred, parallel with a fall in the level of dieldrin in their tissues. The effects of sublethal levels of organochlorines have been confirmed by laboratory experiments which show that they affect the thyroid and the calcium metabolism, with the result that abnormal eggshell development takes place. It seems that, in Britain, we have been fortunate enough to avoid the serious environmental damage which has occurred, with heavier pesticide usage, in the U.S.A. and other countries, but that the margin of safety has not been great. Thus, there is good reason to stop using these persistent chemicals as soon as possible, even if they may so far have done less damage than is often imagined.

As already pointed out, the danger of organochlorine insecticides is that they may be reconcentrated in living systems. One often hears it suggested that this commonly occurs step by step in food chains, thus if the chemical is applied to the soil, it may be taken up by worms, passed on to worm-eating birds, and finally concentrated (perhaps to dangerous levels) in hawks preying on them. This process is probably less important than is often imagined. It does occur, but seldom builds up to anything approaching lethal levels in terrestrial food chains. The real danger is from the few individuals with really high rates of contamination.

In water, however, the situation is different. Fish and shrimps are able to concentrate pesticides either in solution or when absorbed onto minute particles in suspension, by a factor of 10^3 or even of 10^4. Thus they do indeed build up, in their tissues, levels of organochlorines which are harmful and even lethal. This was seen in the Rhine in 1968, when another organochlorine, endosulfan (6,7,8,9,10,10-hexachloro-1,5,5a,6,9,9a-

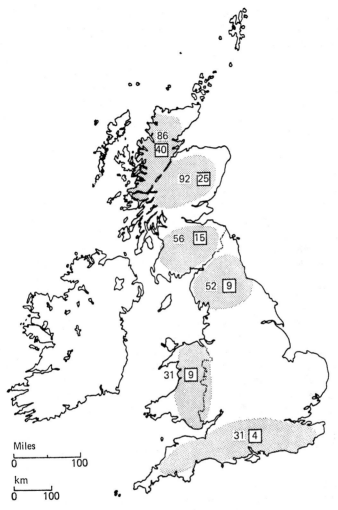

Fig. 7–1   Peregrines in Britain in 1961. In each region the first number shows the percentage of territories occupied, the second (in box) the percentage in which young were reared. (After D. A. Ratcliffe, from MELLANBY, 1967.)

hexahydro-6,9-methano-2,4,3-benzo ⒠ dioxathiepin 3-oxide) was accidentally spilled into the river. Within a few days millions of fish died, though the initial level may have been as low as one part in ten million, a concentration which would have been harmless to man had he drunk the water.

In estuaries and water near the shore, insecticide levels may similarly be high enough to be concentrated by fish and marine invertebrates to significant levels. This view is supported by the finding that many marine birds have substantial levels, and that seals and porpoises from the Baltic have had as much as 55 ppm in their blubber. Though not lethal, this is likely to have had some deleterious effect.

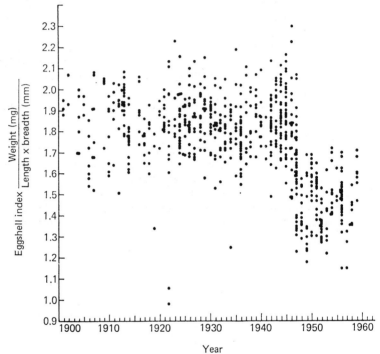

Fig. 7–2   Change in time of eggshell index (relative weight) of the peregrine in Britain. (From Ratcliffe, D. A. (1970). *J. appl. Ecol.*, **7**, 67–115.)

There is no doubt that there is some DDT in the body of our oceans, though the level must be below one part in 10^{12}, as it has not been measured in analyses with this degree of accuracy. Concern has been expressed lest this insecticide should harm the phytoplankton, and upset the oxygen balance of our atmosphere. Experiments have indeed shown that DDT will inhibit photosynthesis of plankton, but only at concentrations some 10^3 or more times those at present existing. As world usage of DDT (and of other organochlorine insecticides) is falling, as is also the level in the tissues of most sea birds, the risk of the oceans of the world being seriously affected by pesticides is slight.

7.6   Recent developments in insect control

During recent years, many pest insects – both carriers of human disease and pests of crops – have become resistant to different insecticides. This resistance, meaning that pest control becomes more difficult, is sometimes thought of as a type of 'pollution'. So far, the pests have eventually been controlled by using a new chemical, but some people fear that a time may come when our armoury of poisons is exhausted and chemical control will then become impossible. This is one reason why many scientists wish to reduce the use of chemical pesticides, or to rely on other methods.

Some pests have been controlled for many years by their natural enemies – 'biological control'. Non-selective pesticides unfortunately also kill these enemies, which are, from man's point of view, beneficial insects. Now attempts are being made to use chemical pesticides more carefully and more skilfully, trying to use natural control when necessary, and chemicals only as a last resort. This is generally described as 'integrated control'.

The result, from the environmental point of view, of all of these developments, and of the production of less toxic pesticides, is that the dangers to the environment foreseen so dramatically by Rachel Carson in *Silent Spring* are now getting less and not more serious. We must still be vigilant, in case unexpected damage occurs, but, in general, widespread environmental pollution from pesticides is probably a thing of the past.

Appendix

A.1 Practical exercises on pollution

I have received many enquiries about possible exercises on pollution, its measurement and the assessment of its effects, from teachers and lecturers who wish to include such work in their curriculum, as well as from members of the public. Unfortunately, work of this kind is seldom simple. Ecological studies, which are the most rewarding, usually take a long time, and require a detailed knowedge of the animals and plants concerned. Thus, students may have heard that lichens are particularly susceptible to the effects of SO_2. When they visit an area where pollution is high, they may be surprised to find a rich growth of such a lichen as the species *Lecanora conizaoides*, which is able to flourish where the more susceptible species are entirely absent. This observation may lead them to learn to identify at least the commoner lichens, and this will allow them to make significant observations, but at the outset few of even the more ecologically minded teachers have this particular expertise.

Ideally, I would wish to be able to suggest practicable means by which the students could themselves make meaningful quantitative measurements of specific pollutants, and then follow up these measurements with biological studies which can be correlated with the chemical and physical observations. This is only possible in a limited number of cases, but sometimes sources of information about specific pollutants already exist, and much can be learned by visiting laboratories or treatment works. However, considerable preparation should precede such visits. It is hoped that the reading of this booklet may contribute to such preparation. Scientists are usually very happy to try to help the *informed* enquirers. They do not react with enthusiasm to the student who writes: 'I have to do a project on pollution. Please send me all the literature.'

This appendix, then, is mainly devoted to suggestions for lines of work which may be profitable. But those interested must be prepared to do a good deal of thinking and preparatory study if their results are to have any significance.

Further information may also be obtained from DELTA (a directory of environmental literature and teaching aids) which may be bought from the Council for Environmental Education at the School of Education, University of Reading, price £2.50. This contains information about study kits. It also deals with books, films, filmstrips, slides, posters, games and workcards on the environment. Many refer to pollution. The directory is produced in a ready-punched, loose-leaf form, and additional sheets will be issued to keep it up to date.

A.2   Air pollution

I have already indicated that the study of air pollution is not simple, partly because the levels of pollutants, even in the worst areas, fluctuate so widely. 'Spot' measurements at a particular time, even if these were easily practicable, would have limited interest, as the observer might miss occasional, and lethal, emissions at much higher levels. Most observations give measurements related to the total emission over a period, often 24 hours, and this may be more significant, though it does not make it possible to quantify the effects of short bursts of high-level pollution in distinction to longer productions of lower levels which will give the same mean figures.

The Warren Spring Laboratory at Stevenage in Hertfordshire is responsible for a cooperative investigation measuring the levels of smoke and SO_2 in some 1500 sites throughout Britain. This network gives far more detailed information than is available for any other country. Most of the stations are operated by local authorities or such organizations as the Central Electricity Generating Board. You should obtain the latest copy of *The Investigation of Air Pollution* from Her Majesty's Stationery Office, or from your local library, which should be able to borrow it for you. Although the Warren Spring Laboratory itself is most helpful to enquirers, do not approach them directly for this sort of generalized information. If your local authority operates a centre, find out who is responsible and contact them, and perhaps a useful visit to the site, together with a lecture on the subject from whoever locally is most concerned, will form a good introduction to the subject. You will obtain advice should you wish to open your own station.

An *Introductory Study Pack on Air Pollution* may be obtained from Messrs Philip Harris Biological Ltd., Oldmixon, Weston-Super-Mare, Avon BS24 9BJ.

Some quite unsophisticated measurements can be of value. In very smoky areas (fortunately now increasingly uncommon) the deposition onto pieces of white cloth may be compared at different sites. Rain may be collected and analysed – the pH in some places, where SO_2 is high, will be affected. Such simple measurements often arouse interest among students, and stimulate those most concerned to take their studies further.

The biological effects of air pollution can be studied in three ways: (1) by surveys of the distribution of sensitive plants which are in fact the assessment of the results of 'experiments' which have already been made; (2) organisms can be introduced into areas where pollution is suspected, and the effects observed; (3) organisms can be exposed (in glasshouses or other confined spaces) to controlled levels of pollutants.

(1) Lichens are known to be particularly susceptible to SO_2. Mosses and liverworts may also be affected. Lichen surveys in and around towns have

proved rewarding, but, as already mentioned, require a good deal of expertise. Before making such surveys you are advised to read some of the original scientific papers on the subject. These include:

GILBERT, O. L. (1968). Bryophytes as indicators of air pollution in the Tyne Valley. *New Phytol.*, **67**, 15.

GILBERT, O. L. (1970). Further studies on the effect of sulphur dioxide on lichens and bryophytes. *New Phytol.*, **69**, 605–27.

GILBERT, O. L. (1970). A biological scale for the estimation of sulphur dioxide pollution. *New Phytol.*, **69**, 529–634.

HAWKSWORTH, D. L. and ROSE, F. (1976). *Lichens as Pollution Monitors.* For details see Further Reading, p. 68.

The best way to start learning to identify lichens is to find someone who can do so, and let him show you the technique. You will also require the book by DUNCAN, U. K. (1970). *Introduction to British Lichens*, T. Buncle and Co. Ltd., Arbroath, 292 pp.

The following useful ideas for surveys have been kindly suggested by Dr Gilbert:

(a) Map the distribution of common wall-top species such as *Parmelia saxatilis* (or even *Parmelia* sp.), *Grimmia pulvinata, Lecanora muralis*, down a suspected gradient of increasing pollution. Then map a common epiphytic lichen, i.e. *Parmelia sulcata* (or *Parmelia* sp.), *Evernia prunastri, Pertusaria* sp., or bryophyte, i.e. *Orthotrichum* sp., down the same gradient. Interpret the results. Does pH have an effect on survival? Does species of tree affect survival?

(b) Map the distribution of *Xanthoria parietina* down a suspected gradient of increasing pollution on:

   (i)   calcareous substrata, i.e. asbestos roofs, concrete, limestone;
  (ii)   eutrophicated sandstone walls;
 (iii)   base of trees;
 (iv)   trunk of trees 1 m above the ground.

Interpret the results.

(c) Collect the psocid *Mesopsocus unipunctatus* from standard trees, i.e. mature free-standing ash trees in the open, down a gradient of increasing pollution. Is there any relationship between the number of individuals per unit area of trunk and the cover of epiphytes? Is there any sign of melanism? (See paper by GILBERT, O. L. (1971). Some indirect effects of air pollution on bark living invertebrates. *J. appl. Ecol.*, **8**, 77–84.)

(d) Simply mapping the distribution of any common sensitive lichen or bryophyte (assemblages would be too hard) onto a 1 :50 000 O.S. map using mapping pins will bring out the distribution of air pollution round a town or factory complex.

There are difficulties in all this. One has to be careful to standardize the habitat examined and of course in an industrial belt there may be no lichens worth mapping, within easy reach of the school or college. Perhaps the students could scan old floras to see what has disappeared and thus realize that pollution and habitat destruction often both act together to impoverish the flora. Students in towns could investigate the distribution of *Lecanora conizaoides* and *Pleurococcus* together with the distribution of psocids on trees.

(*e*) If new sources of pollution are opening up nearby, one can make a tracing of local wall top lichen communities onto clear polythene sheeting and watch them disappear over the years.

(2) Different species and varieties of many plants differ in their susceptibility to air pollutants. Unfortunately, the best 'pictorial atlas' is expensive, and is only obtainable from the U.S.A.

> *Recognition of Air Pollution Injury to Vegetation: a Pictorial Atlas.* Edited by JACOBSON, JAY S. and CLYDE HILL, A. Informative Report No. 1, Air Pollution Control Association, Pittsburgh, Pennsylvania (1970). Price: $15.

Susceptible varieties of plants can be grown in areas where air pollution is suspected. It is advisable to grow the plants in standard soil in pots, for the soil in an area where there has been long-standing pollution may itself be contaminated and this will affect growth. Table 3 shows some of the varieties of cultivated plant known to be susceptible.

Table 3   Varieties of plant known to be susceptible to some types of air pollution.

Plant	Injured by
Gladiolus (particularly var. Snow Prince)	SO_2
Garden pea (e.g. var. Kelvendon Wonder)	SO_2 also Ozone
Zinnia	SO_2
Petunia	Ozone
Tobacco (particularly var. Bel W 3)	Ozone
Gladiolus	Fluoride
Maize	Fluoride
Many coniferous trees	SO_2

In regions such as Lancashire in Britain, where industrial and rural areas are contiguous, mapping coniferous trees may give some indication of high SO_2 levels (trees absent) and lower SO_2 levels where they are present. The black blotches of the tar spot fungus on the leaves of sycamore trees only occur in areas of low SO_2.

(3) Some of the plants listed in Table 3 may be grown in a glasshouse or enclosed with polythene, or better still in a bench-top growth chamber,

and then exposed to a current of air containing a known concentration of SO_2 or other pollutant. A crude experiment, able to demonstrate gross damage, may be made by burning sulphur. The calculation on p. 7 shows that 1.5 mg of sulphur should give a concentration of 1 ppm of SO_2 in 1 m^3 air. However, such a level will not be maintained, and it is probably necessary to burn about 1 g sulphur daily in a small glasshouse to obtain, in a space of perhaps two weeks, noticable lesions on plant leaves. Various experiments on these lines may be devised.

More scientifically, a known volume of air containing an exact level of pollutant should be circulated into the glasshouse. If the polluted air is introduced at a slightly greater pressure than that of the outside atmosphere, reasonably constant conditions can be obtained. Experiments of this kind are obviously only possible in institutions with sophisticated facilities and where constant supervision is available.

A.3   Water pollution

Much of our inland water is obviously and grossly polluted, so work on this subject is less difficult than that on air. Much can be learned from visits to local sewage works, where the staff are often prepared to demonstrate the methods used, including analytical techniques for studying effluents. Before such expeditions, lecturers and teachers without first-hand experience of sewage works should make careful preparation and, if possible a preliminary visit. Excursions may also be possible to the laboratories run by the various River Authorities, where excellent work on the study of pollutants may be in progress. Here again, I must stress the importance of preparation by teachers and students before approaching such authorities, in order to get the most out of a visit.

Water analysis is not difficult. Details may be obtained from two of the Handbooks produced in connection with the International Biological Programme.

GOLTERMAN, H. L. (1969). *Methods of Chemical Analysis of Fresh Waters.* IBP Handbook No. 8. Blackwell Scientific Publications, Oxford, 172 pp.

VOLLENWEIDER, RICHARD A. (1969). *A Manual on Methods of Measuring Primary Production in Aquatic Environments.* IBP Handbook No. 12, Blackwell Scientific Publications, Oxford, 213 pp.

An *Introductory Study Pack on Stream and River Pollution* may be obtained from Messrs Philip Harris (for address see p. 63).

Various experiments with water from streams, ponds and rivers can be made. The following are among those kindly suggested by Professor R. W. Edwards.

(1) Some measure of the growth-promoting properties of waters for algae can be determined using culture of algae (e.g. *Sceletonema* and

Chlorella) incubated under standard light conditions in filtered water samples taken from rivers and lakes (+ dilutions of such waters in a range of standard waters, e.g. distilled or de-ionized).There are many variants to this approach for it can be extended by adding specific nutrients. Growth can be measured by filtering off the algae after a given interval of time (say, one week) and determining chlorophyll concentrations using a colorimeter with appropriate filter – or even by weighing the algal crop.

(2) Toxicity experiments taking samples of effluents or river waters using invertebrates. Cladocerans like *Daphnia* are very good as they only need small beakers, and with planktonic animals one can record the time it takes before they fall to the bottom. These experiments can also be conducted with known concentration of poisons. This is a very good lead into toxicity studies LC_{50} (i.e. the concentration lethal to half the population), slope functions, and so on.

(3) Toxicity experiments in rivers, etc., using small nylon-covered cages anchored to the bed containing invertebrates like *Gammarus*.

Other exercises are described in *River Pollution Studies* by BEST, G. A. and ROSS, S. L. (see Further Reading, p. 68).

A.4 Thermal pollution

If access to water being heated (e.g. by a power station) is possible, temperature measurements can be made, and can be related to studies of the flora and fauna, much as in other investigations of water pollution.

A.5 Radiation

This is not a subject suitable for investigation except by experts under controlled conditions.

A.6 Pollution of the sea

Studies may be made of bacteria near sewage outfalls. The beach flora and fauna of such areas can be compared with unpolluted regions.

A.7 Pesticides

Fortunately, the long-lasting organochlorine pesticides are not available for field tests, so long-term pollution studies with them cannot now be made. Analyses of animal tissues for pesticides are usually outside the capabilities of any but the expert.

Many ecological experiments on plants and animals may still be devised. Thus, broad beans can be given different insecticidal treatments, and the number of aphids and of their natural enemies may be counted. Long-term effects of herbicides on the floral composition of grass sward may be studied. Observations on the numbers of many species of wildlife on farms following different practices and different pesticide used can be rewarding.

Further Reading

ANDERSON, J. W. (1978). *Sulphur in Biology.* Studies in Biology no. 101. Edward Arnold, London, 64 pp.

BEST, G. A. and ROSS, S. L. (1977). *River Pollution Studies.* Liverpool University Press, 92 pp.

CARSON, R. (1963). *Silent Spring.* Hamish Hamilton, London, 304 pp.

COMMITTEE OF THE ROYAL COLLEGE OF PHYSICIANS OF LONDON ON SMOKING AND ATMOSPHERIC POLLUTION (1970). Pitmans Medical and Scientific, London, 80 pp.

FLOOD, M. and GROVE-WHITE, R. (1976). *Nuclear Prospects: A Comment on the Individual, the State and Nuclear Power.* Friends of the Earth, London, 70 pp.

GREEN, M. B. (1976). *Pesticides, Boon or Bane?* Paul Elek, London, 111 pp.

GUNN, D. L. and STEVENS, J. G. R. (Eds) (1976). *Pesticides and Human Welfare.* Oxford University Press, 290 pp.

HAWKSWORTH, D. L. and ROSE, F. (1976). *Lichens as Pollution Monitors.* Studies in Biology no. 66. Edward Arnold, London, 64 pp.

HYNES, H. B. N. (1960). *The Biology of Polluted Waters.* Liverpool University Press, 202 pp.

MELLANBY, K. (1967). *Pesticides and Pollution.* New Naturalist No. 50. Collins, London, 272 pp.

PERRING, F. H. and MELLANBY, K. (Eds) (1977). *Ecological Effects of Pesticides.* Academic Press, London, 204 pp.

ROYAL COMMISSION ON ENVIRONMENTAL POLLUTION. A series of Reports from 1971 onwards. H.M.S.O., London.

SAUNDERS, P. W. J. (1976). *The Estimation of Pollution Damage.* Manchester University Press, 126 pp.

SCORER, R. S. (1973). *Pollution in the Air. Problems, Policies and Priorities.* Routledge and Kegan Paul, London, 148 pp.

SPEDDING, D. J. (1974). *Air Pollution.* Oxford Chemistry Series. Clarendon Press, Oxford, 80 pp.

TAYLOR, F. E. and WEBB, G. A. M. (1978). *Radiation Exposure of the U.K. Population.* R77. National Radiological Protection Board, Harwell, England, 96 pp.